PALAEONTOLOGY OF NITI IN THE NORTHERN HIMALAYA

PALAEONTOLOGY OF NITI IN THE NORTHERN HIMALAYA

BEING DESCRIPTIONS AND FIGURES OF THE PALAEOZOIC AND SECONDARY FOSSILS

RICHARD STRACHEY

GYAN

Published by

GYAN

Gyan Publishing House
5, Ansari Road
Daryaganj, New Delhi-110002
Phone: 011-47034999, 9811692060
E-mail: books@gyanbooks.com

Distribution Network
gyanbooks.com
India, USA, Canada, UK, Australia

ISBN: 978-93-7133-279-8 (PB)

First Published, 1865

2nd Impression 2026

Printed at: Gyan Press, Delhi.

PALAEONTOLOGY OF NITI IN THE NORTHERN HIMALAYA
Author: RICHARD STRACHEY

PALÆONTOLOGY

OF

NITI IN THE NORTHERN HIMALYA:

BEING DESCRIPTIONS AND FIGURES

OF THE

PALÆOZOIC AND SECONDARY FOSSILS

COLLECTED BY

COLONEL RICHARD STRACHEY, R. E.

DESCRIPTIONS BY

J. W. SALTER, F.G.S., A.L.S., AND H. F. BLANFORD, A.R.S.M., F.G.S.

REPRINTED WITH SLIGHT CORRECTIONS FOR PRIVATE CIRCULATION
FROM COLONEL R. STRACHEY'S FORTHCOMING WORK ON THE
PHYSICAL GEOGRAPHY OF THE NORTHERN HIMALYA.

CALCUTTA:

O. T. CUTTER, MILITARY ORPHAN PRESS.

MARCH, MDCCCLXV.

I. DESCRIPTIONS

OF THE

PALÆOZOIC AND SECONDARY FOSSILS.

BY J. W. SALTER, F.G.S., A.L.S.,

OF THE GEOLOGICAL SURVEY OF GREAT BRITAIN,

AND

H. F. BLANFORD, F.G.S.,

ASSOCIATE OF THE SCHOOL OF MINES, AND LATE OF THE INDIAN GEOLOGICAL SURVEY.

---

1. PALÆOZOIC, BY J. W. SALTER.

WHEN, in 1851, Colonel Strachey read a paper before the Geological Society, on the Physical Geography and Geology of the Himalayas and Thibet,* the fossils he had laboured so strenuously to collect hardly drew the attention they deserved. In the published abstract, however, the main facts were recorded, and lists of genera quoted, which showed, in accordance with the sections, an ascending order from Silurian slates, through Carboniferous beds (the Devonian has not yet been detected in India), to a most interesting set of strata, which clearly represent in every way the New Red Sandstone. The fossils of the last-named beds, indeed, agree perfectly with that portion of the Triassic fauna which is developed in the St. Cassian beds, in the Tyrol. It so happened that these celebrated Alpine strata had, a short time before, attracted renewed attention, and their relative age had been canvassed by

* Quart. Geol. Journ., vol. vii., p. 292.

Emmerich, Merian, Von Hauer, and others. This part of the collection was new to Indian geology, and therefore of paramount interest. The oolitic fossils (Lias, Oxford Clay, &c.) were not such novelties, as fossils of that date had been previously known in the Himalaya, and had also been described by Captain Grant from the Cutch district.

With regard to the Palæozoic rocks of India, however, so little was known at the time of Colonel Strachey's researches, that to have secured a fossiliferous base was a great stride in the geology of India. The list of genera furnished by myself to the paper above quoted sufficiently indicated the presence of a lower Silurian group, which, while its fossils agreed in general character with those of Europe, was quite distinct in species. This fact supplies another proof of the existence, at so early a date, of marine natural history provinces like those of the present day. The subdivision of the old ocean fauna, easily recognizable over many areas of Silurian rocks, becomes less conspicuous in the Devonian—the upper part especially, and had become nearly obliterated in Carboniferous times. I shall, perhaps, have the opportunity, in the sequel, of adding a few words on this subject.

From press of time, and numerous engagements, the descriptions will, I fear, be far from complete. But it is, perhaps, hardly to be regretted that every fragment of an imperfect collection should not be described. It is of more use to indicate the genera, and figure the species. Some groups, however, will admit of a little further illustration, and I begin with the Silurian base.

I feel it is due to Colonel Strachey to say, that his share in the working out of the fossils is a large one. The collection was brought home numbered and catalogued, but still required months of patient work in breaking up and chiselling out the specimens. When finally arranged upon tablets, with localities, he placed them all in the colonial collections of the Museum of Practical Geology, and left me the more pleasant task of comparing and describing them. The Silurian species are all of them new.

LOWER SILURIAN ROCKS.

CRUSTACEA.—Trilobites.

There are no less than eight species, all, except one, belonging to ordinary European genera; and the forms are very like those in our own slate rocks. The very first species we notice would alone indicate the Lower- and not the Upper-Silurian rocks; the genus *Asaphus* never being known to rise so high as these, while *Cheirurus, Sphærexochus, Calymene, Illænus,* all of which genera occur in the collection, are as common in the lower as in the upper division. All are essentially Silurian, except *Cheirurus,* which is also known in the Devonian rocks.

1. ASAPHUS EMODI.

Plate 1, figs. 1–9. (Fig. 3, *var. brevior.*)

A. capite brevi. Glabella distincta pyriformis convexa, sulcis utrinque binis exaratis,—anticis obliquis, posticis rectis. Labrum profundè impressum. Pleuræ convexiusculæ. Cauda semicircularis, subplana, axi longo annulato, plano nisi ad apicem convexiori; lateribus 6—sulcatis, sulcis brevibus; margine latissimo.

It is possible that we have two species figured under this name; for the proportions of the glabella in fig. 3 are rather shorter, and the outline of the labrum in fig. 5 is longer than that of the more common one (fig. 4). Nevertheless, as the materials are imperfect, it is not worth while to separate them. If better specimens be found, the short-headed specimen (fig. 3) may then be described as a new species. The specific name is intended to apply first to the caudal shield, figs. 8, 9.

The head must have been rather short, and the glabella convex for the genus, pear-shaped, slightly apiculate in front, and furnished below with two pairs of deep furrows. The upper pair are very oblique, the lower pair (which are really an incomplete nuchal furrow) run direct across. In the shorter variety (or species), fig. 3, the glabella is flatter, and the two pairs of furrows nearly join at their base, which is not the case in the typical variety. The triangular lateral lobes, thus marked out, are themselves flattened, as much so as the glabella, and extend further out than the width of the great frontal lobe, reaching to the base of the eye-lobe itself. The free cheek (fig. 2 *a*) presents nothing peculiar, but seems to have had a very short spine, if any. The surface is smooth.

Pleuræ convex, smooth, with a broad and rather deep furrow. The ends are blunt.

The labrum (fig. 4) is of the usual shape for *Asaphus*, narrowed at the base of insertion, wide in the middle, and with broad lateral lobes, in this species rather blunt. The central space is convex, and the two lateral impressions very deep, each inclosing a tubercle. Fig. 5 shows an impression of the under side, and has the usual strong ridge encircling the central space.

Caudal shield (fig. 9) rather wider than a semicircle, and nearly flat, but more convex towards the apex of the long axis, which reaches nearly to the margin, and is convex towards its end; it is annulated throughout by about nine ribs. The sides are marked by five or six furrows, of which the upper one only is lengthened. The rest stop halfway toward the edge, and abut against a broad smooth marginal belt, which occupies half the lateral spaces.

Young specimens (fig. 8) have the tail of the same shape and proportion as the older ones.

Localities.—Chorhoti Pass. Damchen (figs. 8, 9). . Milam Glacier. (Numbers in Colonel Strachey's Catalogue of Localities—Nos. 926, 1678, 1737, 1743, 1744, 1754, 1766.)

2. ILLÆNUS BRACHYONISCUS.

Plate 1, fig. 12.

The entire length could not have been two inches; there were at least eight, probably more, body rings, very narrow, and but slightly trilobate at the fulcral point, which lies far out. The tail is very gently convex, and scarcely trilobed at all.

It is not certain this does not belong to the next species, but it is remarkably flat, whereas that is convex; and the general habit is so different, that I think I am quite right in separating them.

Locality.—Chorhoti Pass. Damchen. (Nos. 1678, 1743.)

3. ILLÆNUS PUNCTULOSUS.

Plate 2, figs. 10, 11.

I. capite ¼—sphærico, valdè trilobato, punctuoso. Sulci axillares profundi modice remoti, capitis medium haud attingentes. Oculi magni a sulcis subremoti, marginem posticum approximati.

Head a quarter of a sphere, strongly trilobed; the furrows which mark out the glabella lie in deep depressions, and are strong, only moderately distant, and do not reach quite half-way up the head. The large eyes are rather remote from these, and much less than their own length distant from the posterior margin. [The cheeks are wanting, except in fragments.]

The back part of the glabella is convex, almost gibbous, and overhanging—an unusual character. Moreover, the whole surface of the head is strongly punctuated, and over the front portions wavy lines are *closely* set,—parallel to the border. These lines fail on the lateral and hinder regions, over which the puncta are still thickly scattered. They, as well as the striations, show also on the inner cast, but of course more faintly.

The rostral shield (the true hypostome) beneath the front margin of the head is very broad and large, with equidistant striæ, but no puncta. We have not the labrum.

The shape of the head is not at all unlike that of *Ill. Bowmanni.* The projecting back of the glabella distinguishes our species from all British forms, and the large puncta separate it from *I. crassicauda*, of Sweden, a species otherwise not unlike it.

Localities.—Chorhoti Pass; one figured specimen, fig. 10, in a pisolitic iron ore. Gunesgunga? (A similar iron ore is found in British Lower Silurian rocks, Carnarvonshire.) Fig. 11, Milam Glacier. (Nos. 926, 1737, 1739, 1754.)

4. CHEIRURUS MITIS.

Plate 1, figs. 14–18.

C. capite convexo. Glabella valdè distincta, oblonga, anticè paullo projecta, lobo frontali modico, alteris vix latiori, $\frac{1}{3}$ glabellæ longitudinem æquante. Sulci superiores et medii recti, breves, minime arcuati, trans $\frac{1}{3}$ glabellæ tantum exarati; basiles recurvi, lobos rotundos, late intervallo sejunctos, includentes; genæ valdè scrobiculatæ, profunde marginatæ, sulco continuo.

Head nearly semicircular, convex, the glabella very distinct from the cheeks, oblong and parallel-sided, rounded and projecting a little in front. It is strongly lobed,—the frontal lobe nearly a semicircle, very little wider than the rest, and occupying only one-third the length of the glabella. The upper and middle furrows are direct, short, very little arched, and reach but one-third across. The lower are very strong and recurved, and they mark out a pair of spherico-triangular lobes, leaving a wide interspace, on which a projecting tubercle is placed. Cheeks strongly scrobiculate, and deeply margined, the furrow continuous at the angle, which is spinous. Eyes nearly central, very large and prominent, showing lenses (fig. 16*a*) under a magnifier.

The general appearance is much like that of *C. speciosus*, *C. bimucronatus*, *C. insignis*, Barr., &c., from all of which differences of proportion distinguish it. *C. Quenstedti* and *C. Beyrichi*, of Barrande, have also some resemblance, and if the caudal shield (fig. 18), which is found in company with our fossil, belong to it—as is most probable—the last-named species is the nearest ally.

The small caudal shield (fig. 18) has the central lobe convex, but not very markedly separate from the side lobes. It is pyramidal, four-ringed, and runs out between the two terminal lobes. There are four lobes on each side; all are equally blunt and tumid (almost tubercular). The number of these is ten (four lateral on each side, and two terminal), and this number is quite unusual in the genus.

The small labrum (fig. 17) which occurs in plenty with the species, has the usual characters; a blunt termination, convex surface, and obscure concentric furrows.

Localities.—Chorhoti, 17,000 feet. About fifty specimens of various parts occurred in breaking up the blocks from the Chorhoti Pass. Damchen, &c. (Nos. 1678, 1737, 1743, 1741, 1744.)

PROSOPISCUS.—Genus novum.

(*Fam.* Cheiruridæ.)

Prosopiscus. Salter. (πρoooπov, a mask.) Caput transversum; glabella brevis, profunde sulcata, lobis lateralibus validis. Genæ latæ convexæ punctatæ, sine oculis vel suturâ faciali. Corpus ignotum. Cauda (forsan ?) ut in Cheiruro, spinosa.

Head transverse, with deep furrows, a short glabella with strong direct side lobes, broad convex punctate cheeks, without eyes or facial suture. Body unknown. Tail (probably), as in Cheirurus, spinose.

The entire absence of eyes and facial suture is quite enough to distinguish the genus from *Cheirurus.* It is distinguished from *Placoparia* (a Silurian form with minute anterior eyes) by the non-radiating furrows; and from *Typhloniscus,* a genus instituted by myself for a South-African form, also blind and without facial suture, by the shape of the glabella and its non-radiating furrows. Doubtless the lost portions of the body, when found (if any explorers should collect in these snowy passes again), will present corresponding traits of distinction.

5. PROSOPISCUS MIMUS.

Plate 1, figs. 19, 20.

P. capite transverso. Glabella capite brevior, antice angusta, lobis lateralibus tribus compressis,—basali angusto, trigono, convexissimo,—reliquis transversis.

Head transverse, the glabella shorter than its length, and narrowed in front, with a small convex forehead lobe, and three compressed lateral lobes, the lowest most convex and projecting furthest out. This lower lobe is narrow, trigonal—the rest are transverse. The furrows are deep and sharp, the front one shortened, the first (upper) lateral lobe curving round its end.

Between the glabella and the cheeks in front is a small depressed triangular space, and the margin continues all round the head, of moderate thickness, a set of crenulations occurring in the furrow in front of the cheeks. These latter are wide, roughly spherical, triangular, convex and punctate all over, with no trace of an eye or facial suture. The posterior angles are a little produced, but the general form is remarkably oblong and transverse.

Of the tail-piece (fig. 20) supposed to belong to this, we have only the lateral lobes of one side; and no axis. It is probable, but uncertain, that there were only three widely-divergent pleuræ, curved gently, divided for two-thirds their length into spines, and with deep short furrows (not extending to the edge of the limb) between the connected portions of the pleuræ.

Locality.—Chorhoti Pass. Damchen (heads); above Bompras (tail). Nos. 1678, 1737, 1744.)

6. SPHÆREXOCHUS IDIOTES.

Plate 1, figs. 21, 22.

We have no parts of the head but the characteristic basal obes of the glabella, which, however, being round and convex, and widely separated by an arched intervening space, will not allow us to arrange our species in any other genus.

The tail-piece (which can belong to no other) is transverse, formed of at least four rounded subtruncate lobes, of which, as we see the inner or under surface only, it is needless to give any further description. The size corresponds nearly with that of the glabella figured above it, and it is clearly different from the corresponding portion in *Cheirurus mitis*.

Locality.—(Head) above Bompras. (Caudal shield, fig. 22) Chorhoti Pass. Nos. 1663, head; 1737, tail.)

7. LICHAS TIBETANUS.

Plate 1, fig. 23.

L. glabellâ aspero-tuberculatâ, sub-pentagonâ, e tribus lobis magnis composita; lobo centrali oblongo lineari, per medium contracto, ad basin paullo latiori, antice expanso, lateralibus ovato-oblongis, supra latioribus, usque ad basin glabellæ attingentibus; lobis basalibus nullis.

Glabella roughly pentagonal, as long as wide, with three large lobes, of which the central is linear-oblong for two-thirds of its length. It expands above into a rounded forehead lobe, not more convex than the rest of the head. The side lobes are ovate-oblong, rather broadest above, and reaching the very base of the glabella, without any basal lobes between them and the neck border. The rest of the head, body, &c., is unknown. Surface rather roughly tubercular.

The specimen reminds us of several European species, but does not quite agree with any. The basal lobes reach completely to the base in such species as *Lichas Hibernicus* of Portlock, a convex form in the English Lower Silurian.

Locality.—Damchen. Chorhoti Pass, above Bompras. (Nos. 1662, tail; 1678, head; 1737.)

8. CALYMENE NIVALIS.

Plate 1, figs. 24-26.

C. sesqui-uncialis ; capite subtrigono ; caudâ rhomboideâ convexâ ; abdomen deest. Glabella parabolica, brevis, longitudinem ½-capitis æquans ; lobis duobus basalibus intervallo lato separatis ; tertio minuto. Margo anticus latus rotundus, â genis vallo lato sejunctus. Oculi eminentes. Cauda convexissima elongata, axi antice lato, postice contracto, lateribus parallelis, apice abruptâ. Axis sulcis circiter 8-propinquis ; latera 6-costata, costis ad marginem attingentibus, duplicatis. Labrum angustum, oblongum, sulcis duobus concentricis, apice bilobato.

The species is near to several European ones, yet distinct from all. It has the short glabella and prominent eyes of *C. brevicapitata*, and is even more gibbous than that species. The tail has a remarkably produced convex tip, like that of *C. Tristani*, which has, however, fewer ribs, more deeply interlined.

About 1½ inch long, with a subtriangular head and rhomboidal convex tail. We have no body segments.

The glabella is convex, parabolic, short—only measuring half the entire length of the head, with the usual three lateral lobes, the basal ones more than their own width apart; and the other one minute. The front margin broad, rounded, separated by broad depressions from the cheeks. Eyes very prominent, opposite the middle glabella lobe. The tail is very convex; the axis is broad in front, tapers backwards to about the half, and thence is parallel-sided to the end of the prominent blunt tip. The axis is closely scored half-way down by eight or nine ribs; the declining lateral lobes have six stout ribs, which reach the extreme margin, and are duplicate for a short distance from the edge.

The labrum (wrongly called hypostome by many authors) is narrow-oblong, with two concentric furrows and a bilobed tip.

This genus, had no other been found, would have marked the beds in which it occurs as Silurian, for it begins at the Silurian base, and reaches the very top, but is no where known to transgress these limits.

Locality.—Chorhoti. (Nos. 1678, 1741, 1754.)

ANNELIDA.

TENTACULITES.—Sp.

Plate 1, fig. 27.

This species is really so like many others from various parts of the world, that it is scarcely worth while to give it a name. It has the rings, which are tolerably regular, quite faint in some parts of the younger shell, in others stronger, and with a few obscure annular striæ. The adult portion has regular strong rings like those of *T. anglicus*. The form is slowly tapering, the tip curved, and the rings a little oblique; but this character is not so marked as in the *T. supremus*, brought by D. Forbes from the summit of the Andes. If, when we know more of Indian Silurian fossils, the species should prove distinct, it might be named *T. indicus.* I think it is better, in doubtful cases, not to give specific names, but to be content with indicating the genera.

Locality.—Damchen. Chorhoti. (Nos. 1678, 1737.)

2. SERPULITES.—Sp.

Plate 1, fig. 28.

This certainly was the tube of an annelide allied to the forms usually called *Serpulites,* and not merely the cast of a filled-up burrow, such as are common in all muddy sediments. Narrow compressed tubes of this kind are frequent in our own Silurian and other Palæozoic rocks, and I do not know that they have received a separate name, to distinguish them from the thick-edged compressed nacreous tubes which first received the above name, and which very possibly are not annelides at all.

Locality.—Milam Glacier. Chorhoti. (Nos. 933, 1754.)

MOLLUSCA CEPHALOPODA.

These are but few, but they remind us, by their general characters, of those of other Silurian regions. This may certainly be said of the species of the genus *Orthoceras*. The *O. Kemas* is exactly like a common Ludlow rock fossil. The species of Lituites can scarcely be otherwise than from the same Silurian group, but it is right to say there is some doubt of the formation from which the Nautilus was obtained.

It is from a loose fragment of black limestone at Gunesgunga, and may have been derived from younger beds—though this does not seem likely. To avoid mixing it with the other fossils, to which, nevertheless, we think it belongs, it will be described first.

NAUTILUS ? INVOLVENS.

Plate 2, fig. 3.

N. involutus, compressus, anfractibus 3 latis, ad umblicum parvum valdè rotundatis, lateribus planis, angulis externis obtusis, dorso plano. Septa obliqua conferta, lente curva, nisi ad umbilicum abruptè flexa.

The name *Nautilus* is retained for this obscure fossil rather as an indication of its possibly belonging to a newer set of rocks. It is too involute and Nautiloid for any Silurian *Lituites* or allied genus known to me, but is not unlike some of the species of *Discites*, a sub-genus of Nautilus common in carboniferous strata.

Our specimen has a diameter of $2\frac{1}{4}$ inches, and has three compressed whorls—perhaps a minute fourth. The outer whorl is nearly four times as wide as the next succeeding, and is swelled and abruptly rounded near the umbilicus, then flat on the sides, and bluntly angular between these and the back. This is flat, and has a greater width than our crushed specimen would at first indicate. The measures, so far as can be ascertained, would be as follows:—

Section broadly sagittate: its length 15. Breadth at the thickest part, near the umbilicus, 16. Breadth at the inner edge of the umbilicus, i. e., of the preceding whorl, 7. Breadth of the back 12. The surface is destroyed.

The septa are very oblique forwards, bent sharply over the umbilical slope, then sweeping forward in a low curve over the flattened sides, and nearly direct across the back. They are closely placed, very much as in the carboniferous species.

Locality.—(1777) Gunesgunga; in a boulder. Formation somewhat uncertain.

CYRTOCERAS CENTRIFUGUM.

Plate 2, fig. 1.

C. ventricosum, ad latera compressum, conicum vix curvum. Os contractum ?—17 lineas longo—13 lato. Siphunculus externus. Septa concava, ad latera arcuata, nec obliqua, intûs multo confertiora.

Slightly curved, ventricose (contracted toward the last chamber?), the section not much compressed laterally, broad, oval, 17 lines by 13, the siphon quite external. The septa are rather close, and concave; considerably arched backwards (but not oblique) on the sides, and therefore projecting forwards on the ventral and dorsal aspects. Last chamber . . . . ?

This has the aspect of a *Phragmoceras* much more than of a *Cyrtoceras*, and was for some time mistaken for one. But the external position of the siphon decides the genus, for in *Cyrtoceras* it varies from a dorsal to a ventral position, as has been observed by Barrande; while in *Phragmoceras* it is always internal.

The ventricose form (possibly contracted at the mouth, for this is not quite certain, and may be due to imperfect preservation) is found in more than one Welsh species of Lower Silurian date.

The shell is slightly curved, very conical from its rapid tapering, and appears to be contracted towards the last chamber. The section is regularly broad-oval; the siphuncle quite external and rather large. The septa are not oblique, rather close, concave, considerably arched backwards on the sides, and therefore projecting forwards on the ventral and dorsal aspects. They are crowded on the inner side of the curve, and of course wider apart on the outer margin.

The whole fragment is 2 inches long; the last chamber is not preserved in our single specimen, nor have we the pointed apex, which was probably more curved than the older part of the shell. (Fig. 1*a*. is a section of the smaller end.)

Locality.—(1720) Rimkin.

LITUITES IULIFORMIS.

Plate 2, fig. 2.

L. discoidalis, anfractibus 5 fere liberis convolutis, subquadratis, quam profundis ½ latioribus; dorso subplano convexiusculo; lateribus obliquis, ad umbilicum declivibus, angulis externis projectis. Septa simpliciter arcuata, lente concava; sipho internus fere marginalis. Striæ asperæ confertæ.

Discoid, and very much flattened; our solitary specimen, an inch and three-quarters in diameter, having at least 4 or 5 contiguous but not embracing whorls, which are squarish in section,—about two-thirds as deep from back to front as wide; somewhat flattened on the sides sloping to the umbilical margin, very blunt and flat on the back, and gently concave on the inner edge. Surface with undulating sharp unequal striæ of growth. The siphuncle is quite internal, almost touching the inner border. Septa rather distant, in whorls 5 lines deep nearly 2 lines apart, not very convace, moderately arched backwards on the sides, and straight, or only gently curved across the back.

This species puts one in mind greatly of the *L. (Trocholites) ammonius* of Conrad, a Lower Silurian species from North America, and belongs to the same division of the genus. It differs in being much less compressed from back to front.

Locality.—Chorhoti Pass, 18,200 feet high. (No. 1733.)

ORTHOCERAS STRIATISSIMUM.

Plate 2, fig. 4.

O. parvum, vix conicum, striis transversis subtilibus asperis, (nonnullis prominulis) obliquis, utrinque undulatis. Septa—? Siphon ?

Broad-oval in section, very slowly tapering, without rings or any inequalities except very fine transverse striæ. These striæ are oblique from back to front, with a slight double wave on the side, so that the curve is downward from the back to the middle of the side, and thence gently upward again before curving downward over the front. The striæ are fine and elevated threads,—in some specimens every seventh or eighth is more prominent than the rest—in other specimens the prominent ones are less frequent.

In the oblique fine striæ, and slowly-tapering form, this resembles a great many Silurian species. *O. tenuicinctum*, Portlock, a Lower Silurian species from Ireland, is especially like ours. Even the wave of the striations (not correctly represented in our figure) is but a little more decided than is usual in the smoother species. The septa are not plainly seen, unless they are indicated by the stronger rings ; but this is somewhat uncertain. Siphuncle ?

Locality.—Chorhoti Pass, 17,000 feet. (Nos. 1743-4.)

ORTHOCERAS KEMAS.

Plate 2, fig. 5.

O. modicum, subcylindricum, annulis obliquis remotiusculis lineisque longitudinalibus ornatum. Septa ?

A fragment only 1½ inch long, and about one-third of an inch diameter, is all we have of this species, which by its aspect so much recalls several common Silurian shells. The *O. Ibex* of the Silurian System is especially like it, and as there is a Thibetan antelope (the *Kemas*) with horns of the *Ibex* type, I have adopted a specific name suggested by Colonel Strachey.

Diameter 5 lines, very little tapering, and with an oval section. Rings oblique, not waved, and scarcely 2 lines apart, blunt, prominent, not so broad as the spaces between them. Surface much abraded, but showing traces of longitudinal striæ as in the English species to which it is allied. Neither septa nor siphon are visible in our solitary specimen.

Locality.—Kalajowar, 16,000 feet. (No. 1757.)

Fragments of two other smooth species occur. One is $\frac{1}{20}$th of an inch diameter, with somewhat oblique septa and the siphon eccentric. It was found at Upper Rimkin in grit.

The other, from Chorhoti Pass (1744), is doubtful, and may be the young of either of the other species—for it is only $\frac{1}{10}$th of an inch diameter.

THECA LINEOLATA.

Plate 2, fig. 6.

T. recte trigona, dorso curvo, lateribus planis tenuissime striatis. Striæ longitudinales asperæ, æquales,—incrementi nullæ.

About two-thirds of an inch long, and scarcely 2 lines broad, regularly tapering, without any transverse rugæ or striæ of growth, but covered with close sharp longitudinal threads, not wider than the interstices. Section rather more than a right-angled triangle (100° to 110°); the ventral surfaces flat,—the dorsal surface curved (as shown by the section, for the specimen is covered by stone on the dorsal region, and no part of that surface is visible). One specimen only was found.

Locality.—Chorhoti Pass. (No. 1754.)

BELLEROPHON GANESA.

Plate 2, fig. 7.

B. uncialis, discoidalis, umbilico magno, anfractibus 4 apertis, convexis, fere gibbis, prope umbilicum declivem obtusè angulatis. Carina dorsalis prominens, plana. Superficies striato-costata, striis asperis transversis, ad dorsum subrectis, lineisque concentricis obscuris decussatis.

An inch diameter, of 4 very convex whorls, the outer scarcely enveloping the others, which are all displayed in the open umbilicus, giving a discoid or helicoid look to the shell.

The whorls themselves, however, are not compressed, but wide, with a boat-shaped section, steep towards the umbilicus, the sides of which are flat, and are separated from the body of the whorl by a blunt angle. The surface of the whorl is highly convex, and rather inflated, with a sigmoid curve and a slight depression. A furrow occurs on either side of the keel, which is rather broad and prominent, flat or even slightly concave, and quite smooth.

The lines of growth bend much backward, fig. 7 *a*, and are as sharp striæ over the umbilical face, but they become equidistant as sharp ribs (crossed by very fine lines) over the convex sides, and are quite absent on the keel. Near the mouth, which is rather suddenly expanded, but less so than in our fig. 7 *b*, they are closer and more regular. Named after the Hindu god Ganesa (the elephant-headed god).

Locality.—Rather plentiful on the Chorhoti Pass, 16,000 feet. (1754.)

BELLEROPHON.—*Sp.*

Plate 2, fig. 8.

A more globose and much smaller shell than the last, apparently smooth and with a bluntly-angular keel, and the mouth somewhat thickened. It is too imperfect to name.

It resembles much the carboniferous British species, *B. cornu-arietis*, or rather the young state of it, which has been called *B. navicula* by Swerby.

Locality.—Chorhoti Pass, 17,000 feet. (1744.)

GASTEROPODA.

MURCHISONIA HIMALENSIS.

Plate 2, fig. 9.

M. parva vix semiuncialis, brevis, anfractibus 3—4 depressis tricostatis. Costæ acutæ æquidistantes, centralis tricarinata; areæ intermediæ valdè concavæ striatæ. Basis abbreviata.

Depressed, not half an inch long. Spire of few whorls, which are not much oblique, and are ridged by three very strong concentric ridges. The central one, or band, marked by three ribs, forms the central and quite prominent ridge; and the upper and lower keels, at equal distances from the central one, alike prominent, thin, and somewhat wavy. The facets between them are very concave, covered with sharp and tolerably regular lines of growth, which do not bend much backward to the band,—are nearly vertical below it, and retreat but little as they pass over the blunt short base. The space above the upper band is rather broad and quite horizontal. The young whorls do not differ much from the older ones.

Locality.—Chorhoti Pass, 17,000-17,500 feet. (Nos. 1737, 1744.)

MURCHISONIA PAGODA.

Plate 2, fig. 10.

M. semiuncialis, elongata, anfractibus 4—5 sejunctis, valdè angulatis, carinâ centrali et inferiori instructis. Carina subcentralis eminentior, inferior remota projecta, spatio intermedio concavo.

Turreted, of four or five whorls; the spire elongate, the whorls rather oblique, much separated, and sharply angular at the keel, which is a narrow and prominent band, placed but little below the exposed portion of the whorl: the part above the band is slightly convex; the suture not at all appressed; the part below concave to a second prominent keel. The short basal portion is not produced (as it is in many species) or marked by any spiral ridges. Nor are there any on other portions of the whorl. The lines of growth are rather obscurely marked out, and their exact direction is not ascertained. Mouth?—Diameter of the base four-tenths of an inch. Height of shell about six-tenths.

Locality.—Kalajowar, 15,700 feet. (No. 1766.)

PLEUROTOMARIA TURBINATA.

Plate 2, Fig. 11.

P. parvula, brevis, anfractu ultimo subhemispherico insuper planulato seu concavo, subtûs convexo ferè gibbo. Carina superior bicarinata. Superficies lineis concentricis tenuibus,—incrementi retrorsis. Umbilicus perforatus.

A small neat species, much resembling some carboniferous forms. The diameter only a quarter of an inch, the height rather less.

The last whorl only is preserved in our specimen; it is flat, or slightly concave above, and highly convex below, almost gibbous as it turns inwards toward the open umbilicus.

The surface is covered with fine concentric lines, not very distinct; these occur both above and below the double keel, which is concave between its two bounding ridges.

Locality.—Chorhoti Pass, 16,000 feet. (No. 1754.)

RAPHISTOMA EMODI.

Plate 2, fig. 12.

R. modica, lineas 5 lata, discoidea lenticularis, spirâ depressâ pyramidatâ, anfractibus 5, insuper ferè planis, paullo concavis, carinâ submarginali filiformi instructis, subtûs convexiusculis. Striæ incrementi regulares filosæ, valdè retrorsum curvæ.

The shell is discoid-lenticular, 5 lines broad, and only $2\frac{1}{2}$ lines high, of a low pyramidal shape, with 5 whorls and a minute sixth, which only gradually increase in size—the outer whorl being but one-fifth wider than the preceding. The whorls are plane, or only slightly concave near the margin, along which a fine thread or ridge runs, a little way within the edge. The lines of growth are greatly curved backward to the margin, and are very fine, thread-like, and regular.

But little can be seen of the convex underside, which, however, appears to resemble that of many other species.

It is like a *Raphistoma* in its shape, and in the direction of its lines of growth, which run to the extreme edge, crossing *obliquely* the sub-marginal ridge.

But if the direction of the lines were only changed so much as to be at right angles to this ridge, we could not hesitate to call it a *Pleurotomaria*, so closely allied are the extreme forms of that genus to *Raphistoma*. Yet this seems to me no reason for regarding the latter as otherwise than a true generic type. All the *Raphistomæ* have this lenticular shape, and all want the true band; while in *Pleurotomaria* very few species put on the lenticular form, and all have the true broad sinus defined on either edge.

The genera without a band, *Raphistoma* and *Helicotoma* (see "Decades of the Canadian Survey," No. 1, pl. 2,) form a small group, parallel to that of the genera with a distinct band—*Pleurotomaria*, *Murchisonia*, *Hormotoma*.

I think the genus *Raphistoma* a natural one, and therefore do not coincide with the opinion of my friend Mr. Billings, who considers all these forms as sub-genera of *Pleurotomaria*.

Locality.—Chorhoti Pass, 17,000 feet. (No. 1744.)

RAPHISTOMA. .—*Sp.*

Plate 2, fig. 13.

The proportions of the whorls are nearly the same as in the last species, but they are decidedly less flattened, and less acute in section, as seen in the enlarged figure 13 *a*.

Locality.—Chorhoti Pass, 17,000 feet. (No. 1744.)

TROCHONEMA HUMIFUSA.

Plate 2, fig. 14.

Depresso-turbinata, unciam lata, spirâ brevi, anfractibus 5, quadratis, rugoso-costatis, insuper planatis obliquis, deinde verticalibus tricostatis; subtus—? Sutura planulata. Costa superior maxima, remotior; secunda tertiaque approximatæ minores, in spirâ duæ solum visæ. Lineæ incrementi rugosæ, valdè obliquæ (?)

A fine shell, the largest of the *Gasteropods* yet found in these regions, except the *Bellerophon*. It is about an inch in diameter, and rather less in height, the spire being low-turreted and the whorls rising step-like from each other. They are squarish in outline—the suture somewhat flattened; the upper part of each whorl sloping, flat, or nearly so, the sides vertical. At the outer angle the chief of the three ribs is thicker and more remote than the lower two, and all are somewhat rugose, crossed by irregular lines of growth, so that the shell has an antiquated appearance.

Locality.—Kalajowar, 15,700 feet. (No. 1766.)

CYCLONEMA RAMA.

Plate 2, fig. 15.

C. parva vix tres partes unciæ longa, spirata, anfractibus 4–5 rotundis nec angulatis, concentricé sulcatis. Costulæ planatæ, insuper majores. Lineæ incrementi obscuræ, rectæ.

Height about $\frac{3}{4}$ths of an inch, diameter $\frac{1}{2}$ an inch. Whorls five or six (our specimen shows only the three lower ones), rather tumid, regular, rounded, with no flattening at the suture or base, deeply grooved concentrically throughout—the grooves rather irregular, broader and coarser on the upper half, finer and closer below. The ribs are flat, especially in the upper part of the whorl, and crossed by inconspicuous lines of growth and a few strong ones which are direct across the whorl. Mouth rounded (so far as can be seen—the inner margin is concealed by stone).

The genus *Cyclonema* was formed by James Hall to include those species of so-called *Turbo* in the Silurian rocks which had evidently never possessed a shelly operculum, and have besides an oblique and slightly-sinuous edge to the mouth.

Full notes on this genus and on the preceding one—*Trochonema*, will be found in the descriptions of the first decade of the Canadian Survey. See p. 18.

Rama, a Hindù deity.

Locality.—Chorhoti Pass, 16,000. (No. 1754.)

CYCLONEMA SUBTERSULCATA.

Plate 2, fig. 16.

C. modica, 7—8 lineas alta, tantum lata, anfractibus convexis, sub-depressis, ad basin concavis, insuper fere lævibus, subtûs striatis. Striæ asperæ, a lineis, incrementi sinuosis decussatæ.

Rather larger than the last. We have only the two last whorls; the basal one three times the height of the one above it, somewhat depressed above, and flattened below, sufficiently so to be sub-angular toward the basal edge. The suture of the whorls is not at all flattened. All the upper portion of the whorl is smooth, or marked concentrically by very faint striæ. The base is, on the other hand, strongly and deeply ridged by coarse striæ crenulated by the cross passage of strong lines of growth. These lines are slightly sinuous and oblique—as usual in the genus.

The species a good deal resembles the *Cyclonema crebristria* of the British Caradoc rocks, or the *C. bilix* of Hall.

Locality.—Kalajowar, 15,700 feet. (No. 1766.)

HOLOPEA VARICOSA.

Plate 2, fig. 17.

H. depresso-turbinata, lævis, anfractibus 5 convexis, nec ad suturam appressis, lævibus, varicosis. Lineæ incrementi obliquæ, vix sinuosæ. Basis abbreviata, umbilico parvo? Alt. 5 lin. Diam. 5 lin.

Height and breadth about equal—not half an inch. The shell is depressed, turbinate, smooth, of five whorls, and a minute sixth; all moderately convex and somewhat inflated, the upper one of the spire least so. The suture is not appressed in any of the whorls. Each lower whorl is crossed by an obtuse varix in the direction of the lines of growth, which are rather oblique, very slightly sinuous, and very obscurely marked by a few concentric striæ, which do not disturb the general smooth contour.

The whorls increase slowly, each being about half as broad again as the preceding. The base is short, regularly convex, and the lines of growth distinct on it. Umbilicus small, so far as can be seen.

The varicose whorls give the shell a character that recalls many a more modern genus. As I have not observed this in any of the Silurian *Holopeæ*, it may render the genus a little doubtful.

Locality.—Chorhoti Pass, 16,000 feet. (No. 1754.)

HOLOPEA PUMILA.

Plate 2, fig. 18.

H. depresso-turbinata, lævis, anfractibus 5 planatis suturisque indentatis, varicibus nullis. Basis subplana, umbilico magno angulari. Diam. ¼unc.

Similar to the last, but with flatter whorls above and more decidedly angular below. The base is very nearly flat, or gently sloping to a broad angular umbilicus, and the whorls (three or four in number) are obliquely flattened above, indented at the suture, and crossed by somewhat conspicuous close-set lines of growth.

Our figure represents the fossil as rather too much depressed. It is a small species, barely more than a quarter of an inch in diameter.

Locality.—Damchen, 16,000 feet. (No. 1678.)

PATELLA ?

Plate 2, fig. 19.

The genus is very uncertain; it is an obliquely ovate cast of the interior. One of the sides is straight, the umbo very eccentric, and placed near to this straight edge (which is therefore much the steeper side) and at about one-third from the broader rounded end.

Locality.—Chorhoti Pass, 16,000 feet. (No. 1754.)

LAMELLIBRANCHIATA.

It is not difficult to see that the mollusks of this little Silurian group indicate a deposit made in rather deep water. We may judge this from the numerous Brachiopoda, the small size of the Gasteropods, and the few Lamellibranchiate (or ordinary bivalve) shells. And these, too, appear to be deep-water forms—Nucula and its allies being characteristic of the deeper water;—and *Ctenodonta* is the Palæozoic form of Nucula.

CTENODONTA SINUOSA.

Plate 2, fig. 20.

C. modica, ovato-acuta, lente convexa, concentricè imbricata, anticè rotundata, posticè acuta, sinuata, sinu conspicuo. Nates depressæ, subcentrales. Lineæ concentricæ regulares, imbricatæ. Lat. 10 lino; alt. 6½.

Our figure well expresses the contour of this rather large species, which measures at least ten lines broad, and six and a half lines high, not very convex, but most so in the middle of the disk—the beak being small and low, the anterior* side rounder and rather produced; the posterior long, pointed, with a very slightly arched cardinal slope, and a strongly-sinuated ventral line beyond the broadest part. The lines of growth are sharp, imbricating, and nearly equidistant ridges over those parts of the disk which are unworn, and therefore probably extended in the same style over the whole.

The surface is regularly and gently convex, except on the posterior slope, where a low ridge, bounded by shallow furrows,

---

* *Ctenodonta* differs from *Nucula* not only by having an external ligament, but by the posterior instead of the anterior side being largest, as in most other bivalves. In *Nucula* it is the reverse of this—the posterior side is the short one, and the ligament is internal. Leda is like Ctenodonta in shape.

occurs, opposite to the sinus on the ventral margin. This is a character common to several species.

Locality.—Foot of the Niti Pass, 16,000 feet. (No. 1780.)

CTENODONTA.—*Sp.*

Plate 2, fig. 21.

About half an inch wide, and one quarter high, convex, ridged diagonally, and pointed behind. The form is subovate acute, the beak prominent, and the diagonal bluntish ridge contributes to give a triangular look to the entire shell, though it is really subovate. Surface . . . . ? It is too imperfect to name, but is like many another Silurian species.

Locality.—Same as last. (No. 1780.)

CYRTODONTA? IMBRICATULA.

Plate 2, fig. 22.

Transversa, oblonga, à nate omnino laterali usque ad marginem posticum tumida (auriculo minori postico plano) inde per marginem ventralem latusque anticum brevissimum rotundum plana. Striæ concentricæ confertæ filosæ.

Transverse-oblong, tumid along the posterior slope, and flattened elsewhere, with a lateral almost overhanging beak, which, however, is not prominent. The hinge line straight, and with a small square flat ear posteriorly. The surface is concentrically striated, or rather fine-ribbed. The striæ are prominent (thread-like on the posterior ear), and tolerably regular in position throughout, though they appear fasciculate in places.

The anterior side beneath the beak is remarkably flattened, and this is not due, I believe, to pressure. The hinge cannot be seen, and hence the genus (described by Billings) is rather uncertain.

Locality.—Foot of Niti Pass, as before. (No. 1780.)

Fig. 23.—*Var.* COSTULATA.

Var. β Costis concentricis majoribus, angulatis.

We have only the posterior half of a shell with so much coarser ribs than the other, that many would be disposed to consider it distinct. But the shape of the posterior end is nearly the same. And the ribs, though coarser, have much of the same character as those of the species above described. They are angular, prominent ridges, not set quite at equal distances, but nearly so.

Locality.—Same as above.

BRACHIOPODA.

As usual, these are the most abundant shells of the Silurian deposit. Under generic forms familiar to every student of the older rocks, and (although identical with none of them) representing even the common species of Wales and Shropshire, they stamp the formation as accurately as if we could trace the connection of the beds themselves. This coincidence of numerous genera—and, so far as we know, of specific groups peculiar to the Lower Silurian—is very remarkable and satisfactory when such remote districts are compared. It is a clear and convincing proof of the superior influence of the element of time over that of locality in regulating the palæontology of an epoch.

LINGULA KALI.

Plate 3, fig. 1.

L. vix semiuncialis, oblonga, plana nisi per medium depressa, lateribus parallelis fronte quadrato. Apex angulo aperto 40°. Superficies absit.

Not half an inch long, nearly flat (the centre depressed, oblong, with a square front and rather obtuse angular apex.

The shell about once and a half as long as wide; sides parallel (or very gently curved) for two-thirds their length, and thence sloping backwards to the apex with straight edges (the apical angle about 40°). The front is very straight, and angular at the corners. The interior is cosely striated lengthwise over the long triangular central area.

It is very difficult to give characters for species of this genus, even from perfect specimens, that shall distinguish them. Ours is very imperfect. Enough remains, however, to show that it is, like *L. attenuata*, a British species, which, however, is not so flat, has not the centre depressed, and does not show the strong longitudinal striation internally.

Locality.—Kalajowar (15,800) named from the goddess *Kali* (intensè nigra). (No. 1758.)

LINGULA ANCYLOIDES.

Plate 3, fig. 2.

L. parva, tertiam partem unciæ æquans, convexa, regulariter elliptica, striis concentricis confertis lineata, intûs (?).

Smaller than the last, elliptical, regularly convex; length rather more than once and a half the width. Surface crowded with very fine close lines of growth. Interior—?

If the last species was very imperfect, this is still more so. We cannot see the retral or beak end, and the interior is not visible distinctly. The surface is finely striated concentrically, and there is scarcely any trace of radiating lines of any kind.

The little river Ancylus a good deal resembles this elliptic species. The form is rather a common one in the genus: *L. elliptica*, Phill., and *L. mytiloides*, Sby., of the carboniferous rocks, are both of them like it.

Locality.—Damchen (16,500 feet). (No. 1678.)

LEPTÆNA HIMALENSIS.

Plate 3, fig. 4. a-g.

L. semiuncialis, quam semiovali longior, valdè convexa 7-9 costata profundeque striata, auriculis minoribus nec productis. Valva minor profunde concava omnino striata, costulâ unâ in sulco quoque; major ad natem solûm striata, sed costulis binis.

Rather more than half an inch wide, highly convex, semi-oval, but produced in front. All the surface, except the short pointed ears, ribbed by seven or nine obtuse ridges, the middle one largest, reaching from beak to front. The deeper valve (fig. 4) is very convex, the other (4 *d, e*) very concave. The surface of the former is coarsely striated near the beak only, and *two* fine parallel ribs run down each hollow. The coarse striæ cover the concave valve (4 *d*), and there is only *one* strong rib along each hollow in this valve.

Lines of growth inconspicuous. Interiorly, the muscular impressions are deep (4 *a*), the surface rough with tubercles; ovarian ridges obscure.

The commonest of the Indian *Leptœnœ,* and a very elegant one. It is a good deal like the ribbed varieties of our British *L. transversalis;* but the ribs are coarser, and the double rib between the furrows of the larger valve is an unusual character in this small but well-defined genus.

Locality.—Chorhoti Pass (17,000 feet). (No. 1743.)

L. HIMALENSIS. *Var.* TEXTILIS.

Fig. 5.

Var. costulis 7 minus profundis, striis obscuris; lineis incrementi conspicuis.

Although this seems to have opposite characters to those of the species, I am persuaded it is but a variety, with shallower furrows, the outer two being obsolete. It has much less conspicuous striæ, but the lines of growth, which in the ordinary form (fig. 4) are faint even in the concave valve, are here strong and well marked, so as to decussate the longitudinal ribs or striæ. These longitudinal ribs, though fainter in this variety than in the typical form, are nevertheless present, the difference being solely in the ornament.

Such variations are common in all Brachiopod shells: yet ornament is, in truth, a more constant character than form. There can be no fixed rule for discriminating the characters of species.

Locality.—Chorhoti Pass, 17,000 feet. (No. 1741.)

LEPTÆNA REPANDA.

Plate 3, fig. 3.

L. semiuncialis, quam semiovali latior, valdè involuta rugato-undulata, costis seu rugis validis 5-7, haud nates attingentibus. Valva concava lævis, sulcis jugisque quibusdam costulâ unica ornatis.

Fully half an inch wide, highly convex, semi-oval, but with long produced *ears*. Seven very strong folds or ribs undulate the surface, of which the central one is by far the widest and deepest: they reach about half-way from the front to the beak. Coarse and rather obscure striæ run with the ribs, and on the concave valve one strong rib runs along each furrow, and one, a fainter one, along the centre of the ridges. These appear of course on the intaglio as furrows, and our specimen figured is the impression of the outer surface of the concave valve (magnified at 5 *a*).

No transverse striæ show on three specimens.

It is a pretty shell, well characterized by the great width, and extended ears, and especially by the five or seven broad deep undulations which reach half-way from the margin to the beak.

Locality.—Damchen, 16,000 feet. (No. 1678.)

Plate 3, fig. 6.

LEPTÆNA? DETRITA.

L. ? semiuncialis, regulariter convexa, semiovalis, obscure costata et striata, nec rugata. Costæ in valvâ concavâ (solum conservatâ) circiter 13, vix striis profundis intermediis magis conspicuæ, lineisque incrementi undique decussatæ.

Half an inch wide, regularly convex; semi-oval, with an even surface, without ridges. Ribs numerous; on the concave valve are eleven or thirteen principal ones, not much stronger than the three or four coarse striæ which intervene between them. Tolerably strong lines of growth cross these.

The shape is that of a *Leptæna*; but in the absence of strong principal ribs, and of the interior characters, it is difficult if not impossible to decide whether it be *Leptæna* or *Strophomena*. I incline to think it the latter, but the general appearance is that of the genus in which it is provisionally placed.

Locality.—Kalajowar; 15,800 feet. (Nos. 1766-7.)

LEPTÆNA CRATERA.

Plate 4, fig. 1.

L. semiuncialis, radiatim-striata, transversa, valvâ ventrali perconvexa et umbone depressa; dorsali lente concavâ, auriculis haud productis. Superficies striis crebris interlineatis, concentricis fere nullis. Impressiones musculares ventrales, longi, ad angulum 35° divaricati.

Semi-oval, the length half an inch, the width three-quarters of an inch. The ventral valve very convex, the dorsal slightly concave. The surface with fine close radiating ribs.

The ventral valve in full-grown shells is highly convex; the depth of a specimen nine lines long being fully three lines. The old specimens are proportionally more convex than the younger. The greatest convexity is beyond the middle. The shell is pretty regularly convex, the central area only a little abruptly raised beyond the rest, and occasionally with a slight fold on it. The ears are slightly recurved, but not produced at all.

The surface is radiated by fine ribs, which are close, equidistant, and interlined by other and similar ones; and there are a few silky striæ between these, but only near the margin. The muscular impressions (cardinal muscles) are very long, reaching two-thirds down the shell, forked almost from the beak, and thence diverging at 35 degrees. The separate scars are narrow, linear, and obliquely truncate. At the ends a strong crescentic groove, notched on the median line, circumscribes the visceral portion, but is only seen in full-grown specimens.

The dorsal valve is radiated in the same way. It has a slight depression from the beak outwards, corresponding to the elevation in the other valve. It is regularly concave, not abruptly bent in any part. We have not the interior.

This is rather a large and convex species of a genus, which is, so far as I know, confined to the Silurian beds. *L. sericea*, *L. transversalis*, and many other species, resemble it, but none that I know agrees with it quite.

Locality.—Chorhoti Pass (1754), abundant. At Bompras, 16,000 feet high (1663). Milam Glacier (931). Kalajowar; Upper Rimkin; Gunesgunga.

LEPTÆNA NUX.

Plate 4, fig. 2.

L. vix semiuncialis, oblonga,—cardine productiori, valvis involutis, longis, striatis. Valva major gibba cucullata, longa quam lata, umbone obtuso. Valva dorsalis brevior, latior, profunde concava, area magnâ parallelâ.

A small species, nearly half an inch long, sometimes of an oblong, sometimes of a semi-oval shape, not much longer than broad, but apparently so; with the ventral valve very convex, even gibbous in a longitudinal direction,—the dorsal valve deeply and abruptly concave.

The larger valve, which has a greatly curved and overhanging blunt umbo, has the hinge line somewhat broader than the rest of the shell, so as to equal the length of the valve. And the ears, in those specimens which are of an oblong shape, are a little produced. In others of a rounder shape, they are continuous with the shelving sides, which slope away steeply from the central or gibbous portion. This part is occasionally somewhat flattish, especially near the broad blunt umbo, but is generally regularly convex.

The dorsal valve is deeply concave, and much shorter and proportionally wider than the larger valve (decidedly transverse in the more semi-oval (fig. *e*) forms).

This valve is specially concave under the beak; it often presents an abruptly-depressed nucleus, which lies at right angles to the rest of the valve (this is a raised space of course, on the cast, *d*). It is also at right angles to the area of this valve, which is broad and parallel-sided, and has a covered convex deltidium as in other species. Both valves are closely ribbed by equi-distant thread-like striæ regularly interlined and increasing in number, and crossed at intervals by strong lines of growth.

This little species has the strongest analogy with the British species, *L. tenuissime-striata* of the Caradoc rocks, and with the *L. oblonga* of Russia.

Locality.—Kalajowar (15,800 feet); Milam Glacier. (Nos. 933, 1764, 1766.)

STROPHOMENA.—*Rafinesque; King; Davidson; &c.*

The distinctions of these three genera, *Leptæna, Strophomena, Orthis,* which occur in every Silurian formation, are no less easy to determine by the external characters (with some exceptions, as in the case of the shell last described) than bv a study of the interior structure.

Leptæna is always an involute shell; the valves curved strongly, and wrapping the one round the other, so that one shall be very convex, the other greatly concave. With this character are joined the internal ones of elongated muscular impressions, and the lateral teeth of the hinge being confounded with the central boss in the dorsal (or concave) valve, which is known by the term cardinal process. To this process are attached the pair of great cardinal muscles, which open the shell, by pulling on this process—the fulcrum being the hinge and its teeth.

The characters in *Strophomena* have been illustrated by Prof. King and myself, and the genus accepted by our chief authority on Brachiopod shells, Mr. Davidson. In this genus, though a few have somewhat of the bent shell and transverse form of *Leptæna,* the great majority are flat semi-oval shells, as long as broad, and with a hinge line invariably equal to the greatest width of the shell, often a little more. Internally the muscles are of a quadrate form, and much expanded laterally as well as lengthwise; and the cardinal process is very distinct, large and double.

Orthis proper is seldom flat, and generally has one valve much more convex than the other. The hinge line is often not as wide as the shell. The form is compact, not extended. The muscular scars are quadrate, not expanded either lengthwise or sideways in either valve. And there is a single distinct linear cardinal process.

These are tangible characters, and combining, as they do, external and internal peculiarities, offer safe as well as convenient distinctions for a large group of shells—abundant in species, and very much alike in general appearance. Without this clue it is a hopeless labyrinth of variable forms. To all this it may be added that *Leptæna,* as above defined, is restricted to Silurian rocks: *Strophomena* is Silurian and Devonian; while *Orthis* ranges through all the palæozoic formations. *Leptæna* has but few, perhaps twenty or thirty species known. Both the other genera reckon their specific forms by hundreds.

STROPHOMENA TRACHEALIS.

Plate 3, fig. 7.

S. tenuis parva, vix semiuncialis, fronte (non adhuc invento) deflexo; concentricè perrugata, rugis 7, radiatimque striata. Striæ tenuissimæ, nonnullis remotiusculis exceptis.

Shell very thin, half an inch wide, and rather more than half that length, with concentric regular furrows, and finely radiated with close striæ,—every tenth or twelfth of which is much stronger than the rest: rostral portion (all we yct know) concentrically rugate, with about seven equal-sized ridges, which turn but little outwards at the hinge margin. Ventral valve scarcely convex (almost flat) at the beak. Opposite valve regularly concave, with a minute projection at the beak.

It is possible that the valves are reversed in the above description, and the gently-convex valve with its minute rostral projection may be the ventral or enveloping valve. It is impossible to decide, unless the deflected portion of the valve were present. For the same reason we can hardly be sure of the specific distinction between this and the *S. tenuistriata*, of the British Caradoc rocks. But as all the otherf ossils are distinct from European ones, it is reasonable to suppose this flat-valved shell, with very regular rugæ, a distinct form. The species just quoted is rather irregular in its rugæ, and is besides a far larger shell.

Locality.—Chorhoti Pass, 17,000-17,500 feet. (Nos. 1737, 1738, 1743.)

STROPHOMENA.—*Sp.* (like *S. depressa*, Dalm.)

A shell allied to the above, but with much coarser rugæ, has a front precisely in the way of the *S. depressa*, a very common shell in all the slate rocks of Europe. It has six or seven broad rugæ, and the rostral portion is no broader than the deflected front margin.

Locality.—It occurs in various localities: Milam Glacier; Bompras; Chorhoti; Rimkin.

STROPHOMENA CHÆMEROPS.

Plate 3, fig. 8.

S. transversa, (angulis rotundatis) concavo-convexa, rectistriata; striis asperis. Valva dorsalis ad natem depressiuscula; altera nate elevato angusto convexo, dein per discum plana, ad marginem concava Costulæ rectæ, in utrâque valvâ inæquales, majores 12-20, striis rigidis nec lineis tenuibus alternatæ; omnes lineis concentricis minutis, vix sub lente visis, insectæ.

An inch and a quarter wide, and three-quarters long: one valve convex, the other (ventral) concave; valves finely radiated.

Shell considerably wider than long, the hinge line rather short; the sides arched, and the front somewhat straight. Dorsal valve regularly convex, with a very slight depression at the beak. Ventral valve concave towards the front, flat upon the disk, and with a strong narrow convexity at the beak. Striæ numerous, thread-like alternating, some twelve or thirteen stronger than the rest, (in some specimens the intermediate ones are strong also, so that twenty or twenty-five are prominent ridges,) and interlined by smaller thread-like coarse striæ. Lines of growth inconspicuous, but under a lens, minute close concentric striæ are very evident. There are no concentric rugæ.

A fine shell, very common in arenaceous strata.

Locality.—Damchen. Chorhoti Pass; Gunesgunga. (Nos. 1678, 1720, 1737.)

STROPHOMENA UMBRELLA.

Plate 3, fig. 9.

S. uncialis et ultra, tenuis, semicircularis, auriculis acutis productis, valvis involutis, valde (ut in Leptænâ) curvis; ventrali convexâ, altera valde concavâ. Nates minimæ. Valva minor (dorsalis) in medio depressa ut plicata, altera convexa illic elevata. Margo cardinalis corrugata, plicis minutis; superficies costis remotis filiformibus circiter 20, striisque transversis crebris.

An inch wide and seven lines long, thin, semicircular, with somewhat acute ears. One valve (ventral) convex, and enveloping the deeply-concave dorsal valve. Both valves conspicuously radiated by distant ribs, connected by fine transverse lines. The valves are largely undulated lengthwise, and minutely corrugate along the hinge line for some distance on each side the beak. Ventral valve generally convex, with a raised central fold and some lateral fainter undulations, the stronger ribs, about twenty, with shorter intermediate ones; all distinct and thread-like, without intervening striæ, but crossed by numerous very fine web-like concentric lines and some conspicuous ridges of growth. Beaks not at all prominent.

The involute habit of this really beautiful shell is very much that of a *Leptæena*. The corrugations, however, along the hinge border, betray its real affinity, which its expanded form confirms. I have not seen the interior.

Locality.—Damchen (1678). Chorhoti Pass (1737, 1752): Milam Glacier (Nos. 932-3).

STROPHOMENA ARANEA.

Plate 3, fig. 10.

L. uncialis et ultra, semiovalis, angulis rectis, paucicostata, costis remotis; valvâ ventrali convexâ, alterâ planâ. Valva major regulariter convexiuscula, nate minimâ vix elevatâ; dorsalis plana, vix concava, crassa, intûs cicatricibus, dentibus, septo longo, sulcoque concentrico rugosa. Costulæ rectæ, haud striis interlineatæ, remotæ, rete concentrico conferto connexæ.

Width fourteen or fifteen lines, length eleven lines. Semioval, one valve (ventral) convex, the other flat or slightly concave, conspicuously radiated with distant fine ribs. Hinge line equal to the width of the shell, the ears not projecting.

Dorsal valve very flat, and marked by remote and not very prominent ridges, on a flat surface, crossed by strong concentric striæ as conspicuous as the ribs. In the other valve the ribs are interlined by shorter intermediate ones.

Interior of the dorsal valve with strong teeth, the outer ones widely divergent, and with well-marked muscular impressions—the latter covering a circular space, which stretches over half the length of the shell. A thick median plate reaches four-fifths from the beak to the margin, within which, at some little distance, a strong concentric ridge runs all round. The shell is thick, the striæ only show near the margin internally.

Locality.—Damchen. Chorhoti Pass: Milam Glacier. (Nos. 923, 929, in arenaceous grit, 1678, 1722, 1737.)

STROPHOMENA NUBIGENA.

Plate 3, fig. 11.

S. uncialis et ultra, semicircularis, angulis acutis; valvis lente curvis, ventrali convexiore, alterâ plana seu concava, striis confertis subæqualibus filiformibus radiatâ. Striæ concentricæ aut rugæ cardinales nullæ. Cicatrices interni ut Λ divergentes, angusti.

An inch and a quarter wide, and eight lines long; semicircular, with slightly-produced ears, thin; ventral valve gently and regularly convex, the other flat or slightly concave beneath the beak. Both are covered with fine, but not close, radiating striæ, interlined by others equally slender. No conspicuous lines of growth, or rugæ near the hinge; the surface remarkably even and regular. Within, the ventral valve has a pair of remarkably elongated narrow oval muscular impressions, reaching less than half down the shell in a v-shape reversed, like Λ, and widely divergent at their ends by an interval of more than their own breadth. Dorsal valve with a short central ridge, and widely-diverging teeth. The stronger striæ show internally as ribs on the margin.

Locality.—Upper Rimkin. (R.)

STROPHOMENA BISECTA.

Plate 3, fig. 12.

S. semiunciam longa, $\frac{3}{4}$ lata, semiovalis, angulis, rectis. Valva ventralis lente conveva, tenuicostata, striis intermediis, costâque centrali maximâ, striis transversis minutissimis lineisque incrementi rugosis. Valva altera plana striata, nec costis eminentibus. Rugæ car dinales conspicuæ.

Apparently never more than eight lines broad or six long; generally smaller, with rectangular ears and a long semi-oval shape, slightly convex in the ventral and flat in the dorsal valve, radiated with fine ribs, of which the central one is by far the strongest. A small narrow area toward the beak of the convex valve is elevated, otherwise the valve is regularly and gently curved. The ribs or stronger striæ arch a little upwards, interlined by several striæ, which are fine and crossed by conspicuous ridges of growth in most of the varieties. Along the hinge margin there are rugæ directed outwards. The central rib is very strong, dividing the shell into two apparent halves. And all the ribs, both in the plain and corrugated varieties, are nodular or tubercular. Under a lens, minute transverse striæ are visible on all the ridges.

The flat valve (not figured in our plate) differs in the want of the large central ridge and the conspicuous principal ones. The ribs are equally nodose. There is a minute central prominence on the beak of this valve, which appears to be a constant character.

Locality.—Milam ; Bompras ; Chorhoti ; Kalajowar. (Nos. 1662, 1663, 932, 1737-8, 1757.)

STROPHOMENA HALO.

Plate 3, fig. 13.

S. oblonga, unciam lata, 7 lineas longa, angulis rectis, valvâ ventrali ad natem convexâ, deinde planâ, concentrice subundulaât. Margo cardinalis rugosus. Striæ radiales confertæ, minutæ, æquales, lineisque transversis confertissimis scabræ.

One inch wide, seven lines long, rather more than semicircular ; ears not produced ; ventral valve convex at the beak, then nearly flat, obscurely undulated, rugate along the hinge line, and radiated by fine numerous equal striæ crossing the irregular undulations of the surface ; under a lens they are marked by fine close sharp transverse threads.

A pretty species, with the surface concentrically rugose, but only in parts. Along the hinge line there are several very oblique transverse plates. These are stronger than the rugæ of the disk, which are quite irregular and not very well defined. The front may have been turned down as in

L. trachealis, but there are no indications of it in our specimens which may, however, have been young shells.

Locality.—Upper Rimkin; also Milam. (No. 931.)

STROPHOMENA LINEATISSIMA.

Plate 3, fig. 14.

S. uncialis, latè semiovalis, auriculis acutis, margini lato deflexo plicoso, disco plano. Striæ radiales tenuissimæ, majoribus crebris interspersis minutisque transversis ornatæ. Superficies lineis incrementi rugosa, ad marginem etiam antiquata.

One inch wide, seven lines long. Half a broad oval, with pointed ears; the margin turned abruptly down and longitudinally folded in the ventral valve, and bent backward in the dorsal valve; the disk nearly flat, with no concentric rugæ, but with strong close concentric ridges and lines of growth. The radiating striæ are very fine, close, and unequal in size; the larger ones are numerous, and interlined with very fine thread-like striæ, all rough with minute transverse lines in addition to the larger lines of growth.

Locality.—Damchen; Chorhoti; Kalajowar. (Nos. 1757 1765, 1766.)

ORTHIS THAKIL.

Plate 4, figs. 8-13.

O. fere uncialis, costata, valvâ dorsali valde convexâ, alterâ planâ. Costæ rotundæ 18–20 (in nonnullis varietatibus angulares numerosæ 2–3–4 fidæ). Sinus dorsalis medianus nunc distinctus, nunc obscurus.

A conspicuous shell, about an inch broad, and eight lines long, with fourteen strong rounded ribs (in the typical variety, and eighteen or twenty in the extreme varieties. These ribs are trifid or quadrifid for their lower half in the coarse ribbed forms, and divide into numerous fasciculate ribs in the finer-ribbed varieties.

The dorsal valve is highly convex, with a very slight median depression. The ventral valve flat or slightly concave, except the actual beak, which is convex. Hinge line not more than the width of the subquadrate shell.

The above may be considered as the general characters of this shell, which varies as greatly as the *O. flabellulum* of the British Caradoc rocks, and appears to be the exact representative of it. Like that common shell, it has the median line in the convex dorsal valve slightly depressed, and the ventral valve is flat.

The typical forms of both species have few simple rounded ribs, and the extreme varieties have them very numerous and much branched, and the size is about the same.

I shall now describe the varieties, beginning with the simple-ribbed ones. The specific name is suggested by a Himalayan palm, the *Chæmærops Marshiana (Thakil)*.

Var. a.—CONVEXA. Figs. 8, 9.

—Major, costis 14 simplicibus rotundatis.

About fourteen lines broad and nine lines long, subquadrate: the dorsal valve very convex, its ribs, about fourteen, of which the four central ones form a somewhat raised central space, without any marked median depression. Ribs rounded, about their own diameter apart, and with no intermediate ones.

Locality.—Milam Glacier (931), Damchen; Kalajowar.

Var. β—TRIFIDA. Fig. 10.

—Planior ? costis angustis trifidis.

In this variety, which is smaller (and flatter ?) than the last, the ribs are not so broad, but are angular and divided for rather more than half their length into threes: the secondary or lateral ribs being less prominent than the principal ones.

Locality.—Niti Pass (1782, 1766, 1737, 1754); Upper Rimkin; Chorhoti; Kalajowar.

Var. γ.—SUBDIVISA.

Plate 4, figs. 11, 12.

Var. γ.—Planior ? 8-20 costata, costis trifidis, sæpe quadrifidis, ad marginem vix regularibus.

In this variety the division of the ribs is carried further than in the preceding one; the principal ribs, which are narrow and almost angular (not rounded, as in the type variety) being divided into four or five coarse striæ; the central one scarcely preserves a greater prominence than the rest.

The median furrow in the convex dorsal valve is better preserved in this shell than in *Var.* β. The hinge line is of full length, the ventral valve flat or slightly concave, and the size the same as in other varieties.

Locality.—Kalajowar; 15,800 feet (1765, 1766), Chorhoti Pass (1737, 1752), &c., &c.

Var. δ.—STRIATO-COSTATA.

Plate 4, figs. 13, 14.

Var. δ.—Convexior bilobata, costis circiter 24, fere per totum bi- vel tri- fidis.

This variety completes the series for this variable shell; the central furrow is deeper,—the ribs themselves have become more numerous, and not having room for regular trifurcation, they divide into two, three, or four, according to the space, the intermediate ribs starting almost from the beak.

The general character of the shell is preserved even in this extreme variety, which resembles in this respect, as it does also in form, the English fossil *O. flabellulum*, Sow. In both species the type variety has simple rounded ribs, and from this form a series of variations leads us to a shell which would almost be called striate, if its real affinity with the coarse-ribbed shell above described were not easily traceable.

Locality.—Rimkin. Marked (R) in the collection.

ORTHIS TIBETICA.

Plate 4, fig. 5 (4, junior ?).

O. longa, semiuncialis, utrinque convexa, subcarinata, striata. Valva ventralis convexior, nate prominulâ (areâque magnâ), per medium subcarinatâ ; striata, striâ centrali prominula, reliquis fasciculatis. Valva dorsalis, præsertim ad unbonem convexa, areâ centrali elevata, per medium planatos-ulcata, lateribus declivibus. Striæ incrementi inconspicuæ.

A shell which at first sight much resembles the *O. testudinaria* from the Caradoc rocks of Britain. It is rounded, with a hinge line rather shorter than the width of the shell. The ventral and dorsal valves are both convex, and almost as long as wide. The length is eight lines; the width seven lines. The beak of the ventral valve *a* is a good deal raised; and the area is broad for this group of Orthides. Its convexity is not very great, but it rises into a low carina along the centre, which carina is crowned by one of the principal ribs.

Internally, the muscular impressions of the valve are circumscribed, occupy rather a small space, and are not bilobed.

The dorsal valve *b* is also convex near the beak, but is thence flattened down the middle, and the flattening increased till a broad shallow groove, with a decided elevation on either side of it, marks out the central area. The sides or lateral areas slope away from this central portion, and no part of the shell is truly flat. In the variety figured at *c*, which appears also to be distorted by pressure, the central furrow is broader; it is triangular in both cases.

The striæ in either valve are fasciculate; the larger ridges, which reach the beak, being grouped with and interlined by numerous smaller ones, so as to produce a very unequal surface. None, however, is *very* prominent, but the central one in the ventral valve. This is always, as above stated, conspicuous, and in one specimen (fig. *d*) two others are equally prominent. The lines of growth are very obscure. Near the margin the shell is a little antiquated.

Locality.—Chorhoti Pass (1754, 1737); Bompras (C); Rimkin (R). Fig. 4 is a smaller variety.

ORTHIS COMPTA.

Plate 4, fig. 6. (1737, 1768.)

O. minuta, rotundo-transversa, valvâ dorsali convexâ, profunde bisectâ, costis 16–18 rigidis, quarum 2--3 utrinque prope sulcum medianum bifurcatis. Valva ventralis (?).

A small rounded transverse species; the dorsal valve (all we have) convex, but with rather a deep sharp central furrow, which gives it a strongly-bilobed appearance, and with eight or nine sharp ribs on each side, two or three of which on either side of the furrow are forked from about the middle of the shell.

The ribs are pretty regular, about as thick as the interstices, and a little nodular from the crossing of the lines of growth, but not conspicuously so.

The shell has a marked resemblance to the *Orthis Lewisii* of the British Wenlock rocks.

Locality.—Chorhoti Pass (1737); Damchen (1768, 1678, 1744).

ORTHIS MONTICULA.

Plate 4, fig. 7.

O. minuta, convexa. Valva ventralis gibba, costis 22–24 subæqualibus radiata, areâ magnâ triangulari, umbone obtusâ, sinu nullo. Valva dorsalis (?). Long. 2 lin.; lat. 3 lin. Alt. valv. ventr. 1½ lin.

A small species, with a very large area, of a pyramidal shape, and with a blunt umbo. The area is placed at a right angle to the general surface of the ventral valve, which is highly and regularly convex, without any median depression. It is radiated somewhat regularly by about twenty-two or twenty-four rod-like ribs, crossed by squamous striæ of growth. All the ribs do not extend to the apex of the valve. They are somewhat variable in number and thickness, even considerably so,—a few being occasionally more prominent than the rest, and these reach the umbo.

The area is flat, or a very little concave, and the deltidium (*a*) is obscure, but was evidently a narrow one, and probably covered over.

Locality.—Damchen (16,500. Chorhoti Pass, 16,500 to 17,000). (1678, 1754, 1737.)

ORTHIS UNCATA.

Plate 4, fig. 15.

O. subglobularis ! unciam longa et lata (sicut visa oblonga), valvâ ventrali planiore, alterâ convexissima rotunda. Sinus ventralis in juveni nullus, dein profundus subacutus, in linguam longam productus. Sinus dorsalis nullus. Superficies tenuistriata, rugis concentricis lineisque incrementi nullis.

A very curious and unusual form of the genus, more globular than even the gibbous varieties of the *Orthis resupinata* in carboniferous rocks, and with a deep wide sinus produced into a long tongue, which, however, only indents the margin of the upper valve, and does not raise a convex fold on it. The whole shell is remarkably rounded and smoothed, and has far more the appearance of an *Atrypa* or *Athyris* than an *Orthis.*

The ventral valve is only gently convex, with a short hinge line, and a small and very little prominent beak. It has a median depression running from the beak, which increases suddenly in breadth and depth about half-way down the shell, and forms a broad tongue-like projection half an inch long in a full-grown shell. The borders of this sinus are rather abruptly angular. Within, the dorsal lamellæ are very short, and leave an angular space about 30° wide;—the muscular impressions are very slight.

The other valve is very round and gibbous, with a small prominent beak not projecting beyond the short hinge line;

and it is evenly convex over the whole shell, except a narrow keel which runs down the middle line, and which is not always present. The front is deeply indented,—but not raised by the sinus, which is of a broad lanceolate form.

The lines of growth are conspicuous (and somewhat rugged near the front), and there are short interrupted longitudinal striæ (not ribs) covering the whole surface, but they are very obscure in most of the specimens.

Muscular impressions very distinct, rounded, separated by a low but well-marked ridge, and reaching but a short way down the shell.

Locality.—Chorhoti Pass. (1743, 1744.)

ORTHIS.—*Other Species.*

Besides these species, there are several which, though plentiful enough, and presenting characters sufficient to distinguish them from those above described, are not sufficiently perfect to make it worth while to figure or give them names. We must wait (perhaps a long while—for it is not given to every one to geologize in perpetual snow) for better material. Enough has been done to show how nearly allied this Tibetan fauna was to our own in Carodoc times—and yet its specific distinction was complete.

BRYOZOA OR POLYZOA.

The prominent Silurian forms of the group appear to have been precisely of the same nature in the Indian as in the European areas. The genus *Ptilodictya,* under forms so like those of our Caradoc rocks that one might be taken for the other, is the prevalent form in Britain, and, as might be expected, is the first one that occurred to the search of Colonel Strachey. And, as in our own slate rocks, the narrow bifurcating forms and broad foliaceous species are found together.

PTILODICTYA FERREA.

Plate 4, fig. 16.

P. fuciformis, fronde convexa crassa foliosa bifida; apicibus expansis. Cellulis—?

The flat median plate which separates the upper from the under rows of cells, being of course in the line of greatest weakness, is always the one exposed in these fossils. In the present species it is much waved in the direction of the arched lines of growth.

The fronds seem to have been about 2 inches long, and in this length bifurcate twice, spreading out into broad and wavy segments, with leafy expanded ends: the axils between them are obtuse.

There is a narrower variety, if not a distinct species, in these rocks. We have not room to figure it. But it has much the same character on the whole—the irregular wavy surface of the basal plate, and the broad foliose frond, much broader in some specimens than in others. It is in limestone. The larger variety is in a pisolitic or oolitic iron grit, the same in which the *Illænus punctulosus* (pl. 1, fig. 10) is found.

Locality.—(No. 86.)—In iron grit. *(Ganesgunga)*. The narrower variety in limestone (Nos. 1764, 931.)

P. PLUMULA.

Plate 4, figs. 17, 18.

P. uncialis, angusta, ramis ligulatis haud divaricatis; cellulis rhomboideis parvis.

A small species, with strap-shaped branches, not expanded or foliose, and scarcely 2 lines broad, slightly wavy, but nearly straight in direction, and branching at a very acute angle.

The small rhomboidal cells show on a portion of the somewhat convex exterior (*a* in fig. 18). The ridges between them are simple, not elevated into tubercles or spines of any kind.

Locality.—Nos. 1738, 1754, 1767, 1743, 926.

PTILODICTYA—*Sp.*

Pl. 4, fig. 19 (fig. 20, mag.).

On slabs of yellow grit, much weathered, and containing encrinites and small brachiopod shells, are several imperfect casts of a broad leaf-like species. These also show only the flattened surface of the median plate, which is marked as usual by accretion lines.

The fronds must have been convex, and of considerable thickness, as is shown by the cavities left in the stone. The cells were oblong and close set (as at *a*), and a strong calcareous vertical bar seems to have been present in each cell, reaching nearly to its base, occasionally hollow, but generally solid. So at least, I interpret the hollows marked so strongly in fig. 5, and of which the nature is more distinctly seen in the magnified figure (fig. 6.)

Locality.—(1738).

AMORPHOZOA.

The number of fossil forms of this great group has of late greatly augmented, so far as Silurian strata are concerned. And several genera, hitherto considered of doubtful status, have been placed in it. I may instance such genera as *Ischadites* and *Receptaculites*, and particularly the Scotch fossil *Amphispongia*, a genus related to the modern rock-sponge *Grantia*. To these, too, has been added the great pyriform fossil termed *Sphæronites pomum* by Prof. Phillips, but which, on closer examination, turns out to be a sponge of curious compound structure, the exterior covered with mammillar protuberances.

Such a sponge has recently been found by British collectors in Caradoc rocks, and I have proposed the term *Sphærospongia** for it. The same genus occurs, too, among the Thibet

* See Memoirs of the Geol. Survey, Scotland, sheet 32, by Messrs. Geikie, Howell, and Salter, 1861, p. 136.

fossils, and as the genus is only named, not described in the paper quoted below, I will give its characters here. It is, so far as I know, a truly Silurian form, and it is another link between these Indian strata and our own. The Thibet specimens show the structure remarkably well.

SPHÆROSPONGIA.—*New Genus.*

Ovoid or pyriform masses, hollow within, the parietes consisting externally of convex regular hexagonal or rhomboidal mammillæ, with sinuous margins; the sinuations with salient and retreating angles, and often produced into short processes. Internally [in the Devonian species] the walls are strengthened by longitudinal bars or ribs.

It is possible that the smaller species which has rhomboidal areolæ so arranged as to resemble the engine-turning on the back of a watch, may be of another genus allied to *Ischadites.*

This last *(Ischadites)* is a true regular sponge with roots, and has nothing to do with the *Cystideæ,* to which it was formerly supposed to be allied.

I have almost as little doubt that the strange genus *Nidulites,* found in middle Silurian rocks in Britain, is a sponge somewhat allied to these forms. But as that occurs in the form of flattened plates, with honeycomb openings on both sides, it cannot be a very close ally.

SPH. MELLIFLUA.

Plate 5, figs. 4, 5, 6.

An ovoid mass, about one inch and a half long, and an inch broad, *very convex,* covered with long hexagonal areolæ, only slightly convex, and rather less than a line broad, rugose, but without any central mammilla: their edges sinuate,—only minutely so in some cases,—in others, fig. 4 *a,* produced into long processes.

The convexity and size of the areolæ in this striking species, which appears to be common enough in certain beds of the

Himalayan series, readily distinguishes it from the flatter fossil next described, which accompanies it. From the British species before mentioned, the want of a central mammilla will readily enough separate it.

Locality.—(F); also Chorhoti Pass (growing on *Orthis uncata*).

SPH. INOSCULANS.

Plate 5, figs. 7, 8, 9.

A flattened and slightly-convex rounded disk, about an inch and a quarter broad, and covered with small rhomboidal areolæ —the longer diameter of which appears to be transverse to the length, if it may be so called, of the fossil. The arrangement is that of the *Ischadites,* which the fossil much resembles. But the areolæ are not transversely and longitudinally ribbed as in that fossil.

They are only simply convex, and the edges are greatly sinuated, inosculating with those of neighbouring mammillæ in a way which would remind a botanist of the proteiform cells of some water plants. The common *Callitriche verna* of our ponds shows the structure alluded to remarkably well—in the transparent stipules at the base of its leaves.

Locality.—Niti Pass.

CRINOIDEA.

Crinoid stems of several species occur, some an inch across, mostly round; some pentagonal with round canals, others round with quinquefid canals. None perfect enough to be worth description. No Cystidean fragments have been found, and this is somewhat remarkable.

ZOOPHYTA.

Corals, which appear to have been everywhere abundant, though of few species, in Lower Silurian rocks (but which are no where known in the primordial fauna) existed under much the same forms in the Asiatic area as our own. Of the few brought home in this collection, only one or two are sufficiently perfect to name. The largest and commonest is a species of *Chætetes*, or what appears to be so—of a size unusual in Silurian rocks and much more nearly resembling those of the mountain limestone. I have named it after the hoof of the Thibetan ox, the Yak.

CHÆTETES ? YAK.

Plate 5, figs. 1, 2.

C. pedalis, hemispherica, tumida, basi perrugatâ, cellulis fibrosis radiantibus hexagonis, $\frac{1}{60}$, $\frac{1}{100}$ unciæ latis, rectis, continuis.

A large tumid hemispheric coral; sometimes a foot across; which I cannot but regard as a species of *Chætetes*, allied to the Russian carboniferous species, *C. radians*. When broken, however, it does not show the character usually ascribed to the genus, viz.: the open fracture of the tubes. On the contrary, the surface of the long fibrous cells is exceedingly well preserved, and the coral fractures readily along these, exposing the surface of the walls—as in *Favosites*. But I can see no trace whatever of pores.

The mass is hemispherical; the base roughly and concentrically rugose. The fracture exposes radiating straight tubes of a fibrous appearance; and the surface shows scattered clusters of large cells amongst those of ordinary size—the larger ones $\frac{1}{60}$th of an inch diameter, the smaller $\frac{1}{100}$th. The tubes are hexagonal, with somewhat wavy angles (but not produced into projecting tubercles) and show plainly the numerous (gently concave) diaphragms, very regularly placed at about the diameter of the tube apart. They run in paralled bands —at the same horizon all over the coral—and only interrupted at intervals by the varies of growth which are from $\frac{1}{3}$rd of inch to half an inch apart, and very regularly concentric.

The cells on the surface are too much worn to be sure of the character of their edges. They may very probably be smooth (as in our figure).

Fig. 1 *a*, a natural fracture of a large specimen a foot across; 1 *b*, magnified tubes; 1 *c*, magnified surface; *f.* 2, another specimen showing the base.

Locality.—A.—1770.

HELIOLITES DEPAUPERATA.

Pl. 5, fig. 3.

On the base of one of large specimens of *Chætetes* is the impression of a species of *Heliolites*. It is only a fragment, shewing but three or four of the round calices, but enough is present to secure the genus, and assure us the rock in which these corals occur is Silurian—it lies below the more shelly limestones.

The calices are shallow, but distinct at the edge, about a line apart, and rather more than a line in diameter. In two out of the three visible, there are not more than ten septa, which reach inwards fully one-third of the diameter of the diaphragm. This is uneven; at least in our single and badly-preserved impression it appears so.

Locality.—With the above.

Besides these, we have many fragments of millepore corals in masses of limestone (marked D, 1722.) And an obscure branching cup-coral may as well be referred to *Cyathophyllum* as to any other genus. It is from the same bluish shale which encloses the other forms.

I have no more to say on the Silurian fossils of these cold regions. Everything in the collection tends to confirm the view above taken, viz., that there were Silurian Sea provinces tenanted by like but not identical forms at the same epoch. For, assuredly, if palæontological evidence be worth anything, the Himalayan fossils, not one of which is a North European species, are of Caradoc age.

While these pages are passing through the press, Prof. T. C. Oldham, the Director of the Indian Geological Survey, has brought over his collections from the Spiti Pass, 100 miles to the N. West. He finds there, in slates which lie at the bas of the whole section, specimens of a coral which I cannot distinguish from the *Chætetes* last described.

These are followed in ascending order by beds with carboniferous fossils, some of which, indeed, were brought home by Dr. Gerard many years back, and noticed by Dr. Everest in his communication to the Asiatic Society.* The two sections present, indeed, such parallel features, that some notice must be taken of each in the following descriptions.

CARBONIFEROUS.

Neither in the collection made by Colonel Strachey in the Niti Pass, nor in those brought by Gerard or Prof. Oldham from the Spiti, is there more than a very scanty development of the mountain limestone formation. It is not easy, indeed, to distinguish it from the great series which follows, and which includes the *Triassic fauna*. But so far as known, the same genera, and in many cases the same species, are common to India and Northern Europe, a circumstance well worthy of notice. The mountain limestone shells seem to have been cosmopolitan. In Dr. Gerard's collection at Oxford,† the common Productus of our own rocks— *P. semireticulatus* or *P. antiquatus* (for it is equally well-known under both names) is conspicuous, while the commonest fossil in Colonel Strachey's series is the universal *Athyris Roissyi*. It is the same in other Indian localities.

Dr. Fleming's researches in the Punjab, 1848-1852, made, us acquainted with the rich carboniferous strata of the Salt range.‡ Many of the fossils enumerated in his paper (Quart. Jour. Geol. Society, vol. ix., 1853) could not be distinguished

---

* Asiatic Researches of the Bengal Society, vol. xviii.

† By great good fortune a series of Dr. Gerard's original fossils, presented by Captain Gerard to Dr. Buckland, are preserved in the Geological Museum of Oxford. I have had the opportunity of studying these, through the kindness of Prof. Phillips.

‡ Fleming, Journal As. Society, Bengal, 1853. Report of the Geological Structure and Mineral Wealth of the Salt Range, 1854.

from those of our own Yorkshire limestone. And this list was lately extended and improved by the labours of Mr. Thos. Davidson, who re-described, in the Journal of the Geol. Society for 1861, all the species brought home by Fleming, together with another collection made in the same range more recently by Mr. Purdon.

In this useful paper Mr. Davidson enumerates no less than 28 species of Brachiopodous shells alone, and 13 of these are indentical with British fossils.

With these, however, are of course many forms not known in our own area. Prof. King described from Dr. Gerard's collection a fine species of *Strophalosia, S. Gerardi,** a genus more commonly known in Permian rocks. And in the carboniferous limestone of the salt range, more than one species of *Ceratites* is found. These exceptions may be explained by-and-by, when a more minute examination of the rocks shall be completed.

The collection of carboniferous species made by Colonel Strachey is not a large one; and it is to be observed, that the fossils are found in loose blocks, in the Chorhoti Pass, in a locality filled with silurian slate. Colonel Strachey, however, states there are great faults in this series of rocks, some of which may probably have brought down carboniferous strata from a higher level. There is no such ambiguity in the Spiti section.

The following species were collected from a blackish arenaceous limestone. As they have, with two exceptions, been before fully described, I do not think it necessary to waste type and paper by giving fresh descriptions. Some are too imperfect even to name.

From the Niti Pass.	Other Localities.
1. *Productus Purdoni,* Davidson, Quart. Journ. Geol. Soc., vol. xviii., pl. 2, fig. 5—**Pl. 5, fig. 10.**	Salt Range; Punjab.
2. P.—*Flemingii,* Sow. ? (or, *Productus carbonarius,* De Koninck, Monog. Product. and Chonetes)—**Pl. 5, fig. 11.**	Salt Range, also Europe and North America.
3 *Chonetes Vishnu, n. sp.* described below—**Pl. 5, fig. 12.**	
4. *Athyris Roissyi,* Leveillé. See Davidson's paper above cited, pl. 1, fig. 6 (only a small variety, but pretty plentiful here)—**Pl. 5, fig. 13.**	Europe, Asia, North and South America.
5. *Orthis sp.*	

* Trans. of the Palæontograph. Soc. for 1850, pl. 19, f. 6–7.

6. *Aviculopecten hyemalis, n. sp.* described below—**Pl. 5, fig. 14.**	
TO THESE MAY BE ADDED FROM THE SPITI PASS.	
7.* *Productus semireticulatus,* Martin.	Europe, Asia, America, Australia.
8. *Spirifer,* wide species—with two narrow folds in front.	
9.† *Spirifer Moosakhailensis,* Davidson, or a closely-allied species.	Salt Range, in the Punjab.

I should have added to these a large and remarkable Spirifer‡ abundant in Dr. Gerard's Oxford collection, but that Prof. Oldham finds it in the same beds with the Triassic Ammonites ! It is clear we have yet much to learn of Indian Geology. The only two species that need description in the foregoing list are a brachiopod and a lamellibranchiate shell.

1. CHONETES VISHNU.—*N. Sp.*

Pl. 5, fig. 12.

C. semiuncialis, profunde biloba, transversa, auriculis nullis, striis angularibus irregulariter bifurcatis et interlineatis: striis incrementi distinctis.

Transverse, the ears square, not produced. Half an inch broad, and about a quarter long, convex, strongly bilobed—the furrow reaching to the very beak—striate. The striæ are angular, with equal spaces between them : they bifurcate very irregularly at all distances from the margin,—and are wavy and rather nodular,—crossed by somewhat distant coarse lines of growth, and by very numerous close concentric striæ.

Interiorly, lines of small and rather remote puncta are placed in the furrows between the ridges.

This species differs from all allied forms described by De Koninck, in the strong bilobation and the coarseness of the

* Gerard's collection, Oxford Museum. In the Trans. of the Asiatic Society, the Rev. R. Everest mentions *Producta scotica,* and *Orthoceratites.* It is quite probable that both these may occur.

† Prof. Oldham's collection.

‡ I can hardly distinguish it from *S. Keilhavii* of the mountain limestone of Spitzbergen.—See p. 53, woodcut.

striæ. The coarse ribbed species in his work are not flattened, or bilobed. The bilobed species, on the other hand, are fine-ribbed. *C. Hardrensis*, a shell of the same size, but without a furrow, comes nearest our fossil. It is the common British species.

Locality.—Niti Pass.

2. AVICULOPECTEN HYEMALIS.—*N. Sp.*

Plate 5, fig. 14.

A. semiuncialis. Valva profundior convexa, subtrigona, apice acuto, auribus modicis abruptè depressis. Superficies costis 16 eminentibus, equidistantibus, interlineatis.

The left valve of this small species—only half an inch long and as much wide—is orbicular, regularly convex, sub-triangular, and rounded in front. The beak is sharp, and projecting; the ears unequal—the anterior one small with a few striæ; the posterior broad, notched and pointed, and both abruptly sunk. The surface of the shell is regularly radiated by about sixteen strong sharp ribs, equally distant from each other, and interlined regularly by a finer one between each pair.

The ribs are nearly smooth, a little nodular here and there, and crossed by concentric ridges of growth.

It is distinguished from the *Aviculopecten æqualis* (M'Coy), a species described by him in his 'Carboniferous fossils of Ireland,' by the nearly smooth and equal ribs, and the regularly-convex shell. It is also a much larger species.

Locality.—Niti Pass.

TRIASSIC.

The portion of Colonel Strachey's collection which has excited most interest in England is undoubtedly the small but characteristic series of fossils—chiefly *cephalopod* shells—which represent the Upper Triassic group, so well developed in the Austrian Alps. Geologists were, of course, prepared to find a marine fauna occupying the horizon of the great sterile formation of Keuper sandstone, which spreads through Northern Europe; but the discovery of the now well-known St. Cassian and Hallstadt beds in the Tyrolese Alps caused no little surprise, when first the publication of their fossils by Count Münster, Klipstein, and others showed the peculiar nature of this fauna. It was not till the more complete researches by Von Hauer among the fossils of the Dachstein Mountains, and in the Salzkammergut—the latter carried on and published under the auspices of Prince Metternich—that the true horizon of these prolific Alpine limestones was finally ascertained. The fauna was a strangely mixed one; of primary and secondary forms—of *Spirifers* and of numerous *Ammonites*—of *Orthoceratites* among ordinary mesozoic forms of shells.* And all these mixed with the highly-characteristic genus of the Muschelkalk—the *Ceratites*—a fossil very generally believed to be confined to Trias† rocks alone.

The more extended study of these beds by continental geologists has only confirmed the impression they first gave, that their fossils were an intermediate group between the primary and secondary systems of life. It is true that the small globular fossils which could hardly be distinguished from Goniatites prove to be only the young state of Ceratites, and of Ammonites—both genera of secondary rocks—but, then, these ammonites themselves belong to a group—the *Globosi*—nearer to the Goniatite than any other. It is also true that several

---

* *Nerinœa* is a genus which will be readily recognized as mesozoic.

† It was, however, described by Von Buch from cretaceous rocks in the Lebanon, and also in France, Berlin Trans. 1848; and was recognized by myself in the collection of carboniferous limestones brought by Dr. Fleming from the Punjab.—*Quart. Geol. Journal*, vol. ix, p. 193. The localities there are Chederoo, Kafir Kote, &c.

species of Spirifer are known to range upwards into the oolitic rocks, and hence their presence might be more easily accounted for.

The genus *Athyris* (or *Spirigera*, for palæontologists use both names) is, with the rarest exception, a truly palæozoic form. Yet it is found here mixed with *Terebratula* and *Waldheimia* of a very modern aspect. Ammonites of all sizes, from a few inches to three feet in diameter, are the characteristic shells of the deposit. But with them Von Hauer has described very early forms of the *Turrilite, Scaphite,* and even *Baculite* groups, which show a simpler character in the sutures* than any known in more modern rocks—thus reminding us of palæozoic types. Again, the *Orthoceratites* found in these triassic beds present in most of the forms a truly lateral siphuncle like that of the alveolus of the Belemnite—and sometimes even (in *Aulacoceras*) a thickened exterior coat, which must have been formed, one would think, in the way the guard of the *Belemnite* was made, viz., by successive deposits on the outside from the enclosing mantle lobes.

While, therefore, there is a positive intermixture, though less than appears at first sight, of truly palæozoic and secondary forms, there is also in many respects an evident passage from the types that characterize the older to those typical of the newer systems of life.

The Himalaya fossils of this age are but few. But the most striking and common forms among them are the species most characteristic of the same strata in the Alps. I am compelled to this opinion, though I had long thought the species were only closely related to the European forms, by the personal inspection of Professor Suess, of Vienna. His knowledge of the Alpine species is unquestionable, and

---

* Sitzung-Berichte der Kais. Akad. Wissenchaft, Band. 41. Dr. Suess found, and communicated to Prof. Hauer these precious materials,—*Cochloceras, Rhabdoceras,* and *Clydonites.*

he assures me that the "Hallstadt" beds of the Carinthian Alps have a peculiar band of dark slate, tenanted almost exclusively by two fossils. One is the *Halobia (Avicula) Lommelii* of Münster; the other, *Ammonites floridus* of Wülfen. These are the two conspicuous fossil species in the Himalaya series. Mixed with them we also find other European species. *Ammonites Aon*, in one or other of its Protean forms, is, perhaps, as characteristic a fossil as can be selected from the Tyrolese limestones. The *Amm. Ausselanus* and *Amm. coangustatus* are also European species; while no one can mistake the little deformed fossil *Ammonites diffissus* of Hauer. The Tyrol and Himalayan specimens have the minutest points of structure identical. The *Natica subglobulosa* and the two species of *Orthoceras* are identical,—the *Pecten* most probably so.

The Brachiopoda, being probably from deeper water, might perhaps have been expected to show points of resemblance, if there were any communication between the seas in those ancient times. And accordingly the characteristic shells, *Athyris Deslongschampsii*, *A. Strohmeyeri*, the minute but characteristic *Rhynchonella retrocita*, with the *Waldheimia Stoppani*, are all identifiable, and were easily recognized by Professor Suess, who had described these species. After careful examination I am fully disposed to adopt his opinion.*

The same forms, or some of them, occur in the Spiti Pass. Professor Oldham finds in that region a great distinction between the lower mass of strata inclosing the Triassic fauna, and that above it, which is loaded with Oolitic and Liassic types. The Ammonites are of like forms to those of the Alps, though I do not yet recognize any of the species as exactly identical. But with them occurs a most remarkable fossil—a large Spirifer—which Dr. Gerard also brought home in abundance from the same locality, and which I here figure, as it is closely allied to, if not identical with, a shell (*S. Keilhavii*, Von

* Only the names of the genera were given by me in the notes to the paper above quoted.—*Quart. Journ. Geol. Soc.*, vol. vii., p. 305.

Buch) characteristic enough of the mountain limestone in Arctic regions. It is a fine species.

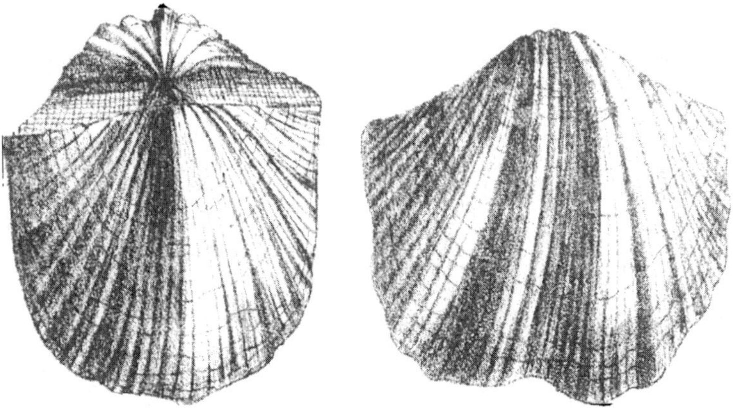

SPIRIFER RAJAH,

n. sp., from the Trias beds, Spiti Pass (Oxford Collection).

S. robustus, subquadratus, longus quum latus, costatus, costis (10-12) acutis multifidis. Sinus angustus, elevatus. Umbo valde curvus.

The peculiar square shape, not very convex valves, acute ribs, are all characters in which *S. Rajah* approaches very near the *S. Keilhavii.* But the fasciculate secondary ribs are stronger, while the costæ themselves are less deep. It is like *S. duplicicosta*, Ph.

It would be strange indeed, seeing that the carboniferous rocks of India are but duplicates of those of Derbyshire; and that the overlying Liassic and Oolitic strata are filled with ordinary British forms, if the intermediate Trias should not contain European species. The identification, therefore, is a very welcome one, and might have been anticipated.

I think, where the species have been already well described and figured by Hauer and other authors, it is unnecessary to re-describe them minutely. The figures will, in these cases, speak for themselves. For the others, the scientific reader may consult the Vienna journals as here quoted.

LIST OF THE SPECIES FROM UPPER TRIASSIC (KEUPER) ROCKS IN THE HIMALAYAH.

Figures.	Upper Trias Fossils.	Species observed in India.	Localities in Europe.
Pl. 6, fig. 1 ...	Ammonites floridus, *Wülfen*	Niti Pass and Spiti Pass.	Hallstadt.
8, 1 ...	" young ♂	Niti Pass.	Carinthia, Hallstadt.
8, 3 ...	" young ♀	"	
7, 6 ...	" Aon, *Münst.* var.	"	St. Cassian.
7, 5 ...	" Winterbottomi, n. sp. ...	"	
6, 3 ...	" sp. (A. Neojurensis *Quenstedt*?)	Niti Pass and Spiti Pass.	St. Cassian?
8, 5, 6	" planodiscus, n. sp.	Niti Pass	
6, 4	" sp.	"	
7, 3	" difflssus, *Hauer*	"	Hallstadt.
7, 7, 8	" Gaytani, *Klipstein*	"	Hallstadt and St. Cassian.
8, 1, 4	" young of A. Gaytani?		
7, 2	" Ausseeanus, *Hauer*	Niti Pass... ...	Aussee, Tyrol.
	" sp.	Spiti Pass	
	" sp.	"	
	" sp.	"	
	" sp.	"	
6, 2	" Blanfordii, n. sp.	Niti Pass	
	Ceratites sp.	Spiti Pass	
	" sp.	"	
	" Jacquemonti, *V. Buch*, 1848	Houkis Pass, E. of Spiti.	
8, 10	Orthoceras pulchellum, *Hauer*? (or O. salinarium, *Hauer*).	Niti Pass, Spiti	Hallstadt.
8, 9	" sp. close septa...	"	
8, 7	" ? flat sp. (Dentalium?)... ...	" ...	sp. like it at Mt. Sandling, Hallstadt, (Prof. Suess.)
8, 12	Natica subglobulosa, *Klipstein*	" ...	St. Cassian.
8, 12	Exogyra, sp.	"	
9, 1, 2	Halobia (Avicula) Lommeli *Wissm* ...	" ...	Salzburg, Carinthia.
9, 3	Pecten scutella, *Hornes?*	Niti Pass and Spiti Pass.	Hallstadt.
9, 4	" sp. minute ribbed	Niti Pass ...	
	" smooth, oblique species, divaricating striæ.	Spiti Pass	
	" sp. smooth, rounded, similar striæ.	"	
9, 5	Lima Stracheyi, n. sp.	Niti Pass ...	The genus Lima is frequent in the Tyrol.
9, 8	Athyris Deslongchampsii, *Suess*	"	
	" (very like Spirigera nux, *Suess*).	Spiti Pass	
9, 10	" (Spirigera) Strohmeyeri, *Suess*	Niti Pass ...	Hornstein, in Lower Austria.
9, 6, 7	Waldheimia Stoppani, *Suess*	"	
9, 7	" do. variety	"	
9, 11	Rhynchonella retrocita, *Suess*	" ...	Aussee.
9, 12	Spirifer Oldhami, n. sp	"	
9, 13	" Stracheyi, n. sp.	"	
Woodcut, (p. 53.)	" Rajah, n. sp.	Spiti Pass (abundant).	
	Lamelliferous cup-corals	Spiti Pass (rare)	

This is not a perfect list, but may serve to show of what materials the Keuper fauna consisted in the Indian area. Suess has beautifully shown how the Red Sandstone areas of this age were grouped round the shores of small Triassic continents and islands in the North European area, while the deep sea deposits between the members of this Archipelago are represented in the south by thick massive limestones and shales, which have, in the course of time, become the highest mountain tops.

We now see that this open water extended continuously to the Alps of India, and the contents of the sea bottom were in the main the same. Yet, around the central masses of India, the equivalents of the New Red Sandstone, as we learn from Oldham's explorations, were again like those of Northern Europe, and they are even more prolific of plants. They were either fresh-water deposits, or deposits close in shore, while the Alpine limestones were formed in deeper water.

Most of the following were found at Raj-hoti; where any of them are from Gunesgunga only, they are so marked.

AMMONITES FLORIDUS.—*Wülfen.*

Plate 6, fig. 1 (Plate 8, fig. 1-3, *juniores*).

Von Hauer. Naturwiss. Abhandl., vol. i., t. 1, figs. 5-14.

If not quite perfect, there is enough to show that this fine species is identical with the common fossil from the Carinthian Alps.

The name has obvious reference to the richly-foliated sutures which show all over the surface in the adult shell. The species having been well described, I shall only notice a few of the principal features.

The curious short ridges, with indents between them, are quite characteristic. The surface looks as if the shell were pinched up at regular intervals along the middle line of the whorl.

Comparing the sutures with those figured by Prof. Hauer, I observe a little difference in the dorsal lobe of our Himalayah specimens, which has rather a long process, intermediate between it and the first lateral lobe. This process is absent in Von Hauer's excellent figure (fig. 14 *f*).

Prof. Suess detected the young of this Ammonite (which by the bye, was divided by its first describer, Wülfen, into four species*), in the Ceratite-like fossils figured in Pl. 8. These young specimens do not in all respects accord with Von Hauer's figure. They are not so thick in proportion; even the fig. 3, which must be regarded as the ♀ form, has not the breadth and obtuseness of Von Hauer's, fig. 6. Nor do his figures show the rudimentary ribs, so conspicuous in Sowerby's drawing (fig. 1*a*, *b*, *d*). This last must be the ♂ form, as it is so much thinner.

These rudimentary ribs just show themselves a little in the youngest portion of our principal figure, Pl. 6. And that M. Suess is right in referring the young specimens to the same form, I do not doubt.

Prof. Hauer considers the *A. Floridus* to belong to the *Falciferi*. Most of the Trias ammonites are of the *Globosi* group.

AMMONITES AON.—*Münst.*

Plate 7, fig. 6.

Münster and Wissmann. Beitr., vol. iv., t. 15, f. 27. Von Hauer. Naturwiss. Abh., vol. iii., t. 5. f. 4-6.

Our specimen is so much unlike Münster's larger figures (l. c. fig. 27) in having only five rows of knots instead of nine on each side, and still less like Von Hauer's of the same species, from Aussee, that I can only yield to M. Suess' judgment, as I have every right to do. The specimen here figured is clearly a young shell, the sutures (fig. 6 *e*) not having lost their Ceratite character.

---

* *Nautilus floridus*, *bisulcatus*, &c. Cärnthnerisch. Helminthol. 1793. It is the *Amm. bipunctulus*, Quenstedt.

Ammon. Brotheus, Münst. l. c. fig. 28, itself, probably, only a young shell, very much resembles our species.

AMMONITES WINTERBOTTOMI.—*N. Sp.*

Plate 7, fig. 5.

Compressed: at the umbilicus it is not thicker than on the disk; diameter, 2 inches: thickness, ½ an inch. The back is rounded, smooth, and neither keeled or ribbed. Umbilicus large, open? Our figure represents it so, but it is rather broken in the specimen, and may have been much more closed. Ribs coarse, not much curved or bent, directed forwards. They are but faint for the inner half, where they form a tubercle, and thence pass, with a very slight bend backward, to the more prominent marginal tubercle. Here and there one is branched from the first tubercle, but this is very rare.

It need not be compared with such sub-globose species as *A. Ramsaueri* (Hauer), in which the umbilical margin is the thickest part of a thoroughly involute shell. I would, however, compare it with Klipstein's *A. æquinodosus*, (Ostl. Alp. t. 7, fig. 1), and it is possibly still nearer to his *A. Mandelslohi* (t. 6, fig. 2). This last is, evidently, a near ally, but the figure is a bad one. The ribs are more branched and are continued beyond the outer tubercle, while the back of our species (fig. *c*) is smooth, or only with fine concentric striæ.

A. bicrenatus, Hauer, is a yet more remote ally.

AMMONITES PLANODISCUS.—*N. Sp.*

Plate 8, figs. 5, 6.

Diameter, nine-tenths of an inch; breadth of last whorl, four-tenths; thickness, two-tenths. Whorls much flattened one-third embracing each other, with bluntish round-keeled back, and no ribs or indentations; section ovate-sagittate. [Sutures? in the young, fig. 6, Ceratite-like.]

Closely allied to *Amm. coangustatus*, Hauer (Sitz. Berichte, vol. xli., t. 5, fig. 1). But that species is less compressed, has a less number of whorls, and the last whorl increases less rapidly than ours.

A. Morloti, Hauer, Naturw. Abh., vol. iii., pl. 2, has a steeper umbilicus.

AMMONITES.—*Sp.*

Plate 6, fig. 3.

A fragment of another smooth discoid ammonite, with very simple foliated sutures. It can hardly be *A. Neojurensis*, Quenstedt (*A. debilis*, Von H.), a species with more rapidly increasing whorls and complicated sutures. But I see no other species to compare it with, and Prof. Suess suggested this one. It is a little like *A. Morloti*, V. Hauer.

AMMONITES.—*Sp.*

Plate 6, fig. 4.

Another fragment of a large discoid shell, with simple serratodentate lobes, much like those of other triassic ammonites.

The simplicity of the sutures recalls those of the lias ammonites figured by Quenstedt as *A. Conybeari*, *A. Turneri*, &c., &c.

Locality.—Gunesgunga.

Fig. 4, pl. 8, is the young of some one of the foregoing Ammonites.

AMMONITES DIFFISSUS.—*Hauer.*

Plate 7, fig. 3 (fig. 1 ?).

Von Hauer. Sitzung-Berichte, Kais. Akad., vol. xli., t. 4, f. 11–13.

Little need be added to Von Hauer's description and figure of this curiously-deformed ammonite, one of a group not at all uncommon in the trias beds. One of the most remarkable of them—with long spout-like contracted mouth, and thick varices—is the *A. delphinocephalus*, V. Hauer; *A. bicornis* and *A. semiglobosus* are other forms of the like character. Some of them have been made into a distinct genus, *Clydonites*, by Von Hauer.

The simple septa made us at first believe we had the young of a larger species. The striated coat is an epidermis, and does not belong to the actual shell.

Our largest specimens are a good half-inch wide, and seven-tenths across the mouth.

AMMONITES AUSSEEANUS.—*V. Hauer?*

Young. Plate 7, fig. 2.

V. Hauer. Naturwiss. Abhandl., vol. i., pl. 8, f. 6-8.

If this be a correct reference, it can only be the young state with the usual Goniatite-lobes. The adult state is a globular mass, some inches across.

AMMONITES.—*Sp.*

Plate 7, fig. 4, and plate 8, fig. 1.

It must be the young state of one of the following species.

AMMONITES GAYTANI.—*Klipstein.*

Plate 7, figs. 7, 8.

Klipstein, Geol. Ost. Alp., t. 5, f. 4. F. Von Hauer, Naturw. Abh., vol. iii., t. 4, f. 13, 14.

It is possible that one of these may be the *A. Johannis-Austriæ*, Klipst. But I think not, the lobes are too simple, and ours are not very young shells.

The striation of the surface is hardly distinct enough to judge by. But it is stronger than is figured in *A. Gaytani*, and not so strong as that given by Klipstein in the royal species.

I understand from my friend Prof. Suess that Von Hauer proposes the name *Arcestites* for this group of globular smooth ammonites with numerous foliated lobes.

All success to the naturalist who first shall rightly divide the Ammonites. A more unwieldy group than it was before Von Buch hit upon the divisions which bear his name could hardly be found, unless it were the Terebratulæ before Dalman attacked them.

The group is still unwieldy, and, what is worse, the subdivisions which answered fifteen or twenty years ago refuse to receive the newcomers. *Arcestites*, *Clydonites*, and any other well-formed *-ites* will be acceptable.

Amm. Aon, *A. Gaytani*, and *A. Johannis-Austriæ* are common to the Hallstadt beds and the celebrated St. Cassian strata.

AMMONITES BLANFORDII.—*New Species.*

Plate 6, fig. 2.

Diameter, fully $3\frac{1}{2}$ inches. Length of oval mouth, $1\frac{8}{10}$ inch. Width, within the middle of the whorl, 1 inch.

Whorls but slightly enveloping, flattened on the sides, quite round on the blunt periphery; not excavated, or steep, or much curved at the umbilicus, and ribbed from it to near the back by straight obtuse ribs, which grow fainter, and do not end in a tubercle or ridge towards the circumference. They are there about one-third of an inch apart. None are intermediate or shorter than the rest. There is a very slight sigmoid curve in each rib, but to all appearance they are straight.

The lobes are three on each side, besides the obtusely-pointed dorsal lobe, which is crenulate—the siphonal interspace (saddle) acute. The side-saddles are strictly rounded and smooth as in Ceratites—the lobes only serrate-dentate—and but simply so—for half their length.

Comparing it with such Ceratites (?) as are figured in Münster's plate 14, we find these ornate species have but two lateral lobes in Münster's figure; but Klipstein's have three; and this is probably nearer the truth.

Comparing with other true Ceratites, we find *C. parcus*, Von Buch, is very like in the sutures, but has no ribs on the whorls; while *C. nodosus*, also from the Trias, has branched and even knotted ribs, and an extra lobe.

Ammonites robustus, Von Hauer (Denkschr., vol. ix., pl. 2), is a species very like ours, but cannot be the same. Ours has closer, and straighter, and continuous ribs, compared in parts of equal size and age with *A. robustus*. Nor are any of the ribs of our species intermediate or broken, as in Hauer's plate 3, fig. 1.

Is it not rather strange that Quenstedt, in his admirable treatise, should have placed the "Globosi" so far from the true Goniatites and Ceratites? The analogy seems much the closest with these.

ORTHOCERAS.

Plate 8, figs. 7–10.

Of the three species here figured, not much can be said, for even a perfect Orthoceras is hard to identify.

O. pulchellum, Von Hauer (Naturwiss. Abh., vol. iii., pl. 1) is perhaps the nearest in form, and in the distance of the septa from one another, to our figures 8, 10. But *O. salinarium* of the same author is also much like, and only seems to differ in the transverse lines which ornament it. (Ceph. Salzk., t. 11.) To one or other of these, the former in preference, I would refer the above figures.

But fig. 9 has closer septa, and must be a distinct species.

And fig. 10 represents a small oval species, which Prof. Suess assures me is extremely like one from Mount Sandling in the Hallstadt beds. I have for the present declined to name it. There are no septa visible, and it might be a Dentalium.

NATICA SUBGLOBULOSA.—*Klipstein.*

Plate 8, fig. 12.

[Ostl. Alpen., plate 13, fig. 13. Compare Nerita Klipsteinii; *Hornes*, Denkschrift., Kais. Akad., vol. ix., plate 2, fig. 7.]

A minute but well-formed shell, with a very small spire, and a rather full suture, i. e. not appressed—the lines of growth a little antiquated near it; a broad ovate mouth pointed above, but blunt below, and no callosity on the inner lip.

In all these characters it agrees better with Klipstein's figure above quoted than with any of the other numerous species figured by him, or the few given by Münster.

It was referred by Professor Suess to *Nerita Klipsteinii*, but I do not know anything of that species, except from Hauer's figure. *N. Klipsteinii* has, indeed, oblique striæ, but I prefer the above identification. *N. Münsterii*, Hornes, is nearer in shape, but still has a callous lip. Ours is probably only a young specimen.

MONOTIS (HALOBIA) LOMMELI.—*Wissmann.*

Plate 9, figs. 1, 2.

H. Lommeli, Wissm. in Münst. Beitr., vol. iv., plate 16, fig. 11. Von Hauer in Denkschr., vol. ix., plate 2, fig. 17.

This shell is positively identical with that figured so well by the above authors. I do not see why it should not belong to the genus *Monotis*. To the *M. salinaria* which sometimes takes its place, it bears a strong resemblance, and the accidental enlargement of the anterior side is surely not enough to give generic character.

Locality.—Gunesgunga chiefly. The small specimens are from Rajhoti, at which place most of the other species above described are found.

The preponderance of this shell, and of the *Amm. floridus* in a peculiar band of black slate among the limestones near

Bleiberg has been specially noted by the Austrian palæontologists. The black slate is almost exclusively occupied by these two shells.

And this fact is doubly interesting when we find that in the remote Himalayas the same two species are the characteristic shells of the blackish limestone. *Halobia Lommeli* also occurs in limestone in the Salzkammergut country, forty miles from Carinthia. In some localities *Monotis salinaria*, as above stated, takes its place.

PECTEN SCUTELLA.—*Hornes?*

Plate 9, fig. 3.

Hauer, in Denkschrift, vol. ix., plate 2, fig. 23.

The ears of our species are more equal; else I do not see why it may not be identified. Ours has fine concentric lines of growth, and very faint radiating lines diverging from the central portion in the curved form so common in Jurassic Pectens.

It is perhaps distinct, being a little too oval, but it may be at present regarded as a variety.

PECTEN.—*Sp.*

Plate 9, fig. 4.

A very pretty, small species, which I dare not name. But I see none in the works above quoted that could be accurately referred to it.

LIMA.

Plate 9, fig. 5.

Half an inch long, and 3½ lines broad, regularly convex, and especially so at the beak, and not flattened or hollowed out on either side. The surface regularly radiated by about 26 flat ribs scarcely raised above the smooth surface, and only divided by impressed lines. Lines of growth inconspicuous.

L. Ramsaueri of Hauer has too many ribes, and *Pecten ? Puchsi* of Hornes is too flat, and has, besides, a different shape. I would name the Himalayan species, if it were more perfect.

EXOGYRA.—*Sp.*

Plate 8, fig. 11.

I think my friend and master, Sowerby, has put a muscle too many in this shell. I cannot find the traces of two in the valve figured (c), but the general form is faithfully given. The deep valve has but a very small beak and attachment, nor is the anterior side much angulated or steep. The shallow valve is gently curved. The shell is smooth, but the characters of the surface are much obscured.

None of the Ostreoid shells figured by Klipstein resemble this. But I hardly know how far the *Ostrea aviculoides* of his plate 15 may be allied to it. That is certainly a distinct species. But these Ostreidæ are difficult to determine in the best of cases, and in this case the specimens are neither full grown, well preserved, or worth description.

BRACHIOPODA.

The shells of this group, from the Austrian Alps, have been fully described by Professor E. Suess. I availed myself, therefore, of his visit to England, and he was kind enough to determine for me the following species, so far as they are identical with European species. I have no need to describe these afresh, but refer to the excellent figures of his monograph, as quoted below.

ATHYRIS (SPIRIGERA) STROHMEYERI.—*Suess.*

Plate 9, fig. 10.

Denkschrift., Akad. Kais. Kongl., vol. ix., t. 1, figs, 4, 5, 6.

Our specimens are a little broader than the Alpine forms, and show nothing of the central furrow. But only one of Professor Suess' does so.

That *Athyris*, a Palæozoic form, is not so unusual a visitor to the upper regions, I learn from my friend, Mr. Davidson, who writes me that a giant species, *A. oxycolpos*, occurs in the overlying Kössen strata, in the Alps.

ATH. (S.) DESLONGSCHAMPSII.—*Suess?*

Plate 9, fig. 8.

Spirigera, Suess in Denkschrift., ib., fig. 3.

Ours is a young shell, but not unlike in shape to the species which Professor Suess has somewhat doubtfully referred to it.

Our fig. 8 shows less of the typical character than two other specimens, which have the front more raised. Austrian specimens of *S. Deslongschampsii* have a deeper central furrow in both valves.

RHYNCHONELLA RETROCITA.—*Suess.*

Plate 9, fig. 11.

Denksch., ib., t. i. fig. 10.

The general boat-like form is very like Professor Suess' figure, but the species could hardly have been identified, except by himself. He has figured a smooth variety. Ours is a plicated from—about five plications showing on each side, and six on the dorsal fold near the margin only.

Locality.—Gunesgunga.

WALDHEIMIA STOPPANI.—*Suess.*

Plate 9, figs. 6, 7.

Suess, in Stoppani's Palæontologie Lombardi (1859), plate 23, figs. 12-16.

I know nothing of this, except from the identification of Professor Suess. But it agrees well with the figure in the Abbé Stoppani's work, as quoted above.

SPIRIFER OLDHAMI.—*N. Sp.*

Plate 9, fig. 12.

A globular species, 7 lines long, 6½ wide, and about as deep, with the valves united as they are in our solitary decorticated specimen.

The species belongs to the group of short-hinged and broad-plaited Spirifers, familiar to us in the Lias. *S. rostratus* is the nearest British type. The beak is blunt and projecting, and somewhat incurved; the area small and short; the sides very steep. The larger valve has a shallow sinus, and traces of about three blunt ribs on each side. The dorsal valve has a blunt, distinct convex fold, and about three ribs on either side, but they do not reach the margin, which, in this old specimen, has coarse antiquated ridges of growth. There are several concentric ridges over the body of the shell, as seen in our fig. 12 *b*.

SPIRIFER STRACHEYI.—*N. Sp.*

Plate 9, fig. 13.

A wider shell than our Lias Spirifer, *S. Walcotii*, with a wider area, and more of a carboniferous aspect.

We have three specimens. The largest is one inch wide, and half an inch high, while the depth of the valves united is scarcely half an inch.

The beak is very little incurved, overhanging a large and wide concave area, and the larger valve has a broad, deep sinus, with angular ribs bounding it, but none on the sinus. Five distinct and rather prominent ribs occur on each side, the *pointed* angles being destitute of ribs (fig. *b*). The other valve (fig. *a*) has a corresponding broad smooth fold, rather convex, and distinctly bounded by furrows. It increases rapidly in size, from the beak to the front margin, so that the side ribs, six on each flank, appear to diverge widely.

A few rugæ of growth occur at regular intervals, crossing these ribs, and the margin is antiquated by them. We have not the outer surface of this shell, which is more carboniferous in its appearance than any other of the Brachiopods above described.

I must not omit the notice the large *Spirifer Rajah*, from Spiti Pass. It is possibly S. *Keilhavii*,* but that is uncertain. What is certain is, that it belongs to a carboniferous group of species, and is probably more allied to the *S. duplicicosta*, Phillips, than might, at first sight, be imagined. I have described it, and it is figured in the woodcut above (p. 53). But it is one of the fossils of the Sipiti Pass, and I have merely quoted the other Triassic forms from that district, leaving their description to Professor Oldham and his corps of scientific men.

CORALS.

A few lamelliferous corals were found in the Spiti Pass, among Ammonites evidently of this formation.

What is singular is, that scarcely one of the Triassic fossils obtained from the Spiti district, and brought home last year by Professor Oldham, correspond with those of the Niti Pass, though 100 miles nearer to the Alps. I could hardly identify a single form with Hauer's figures, and I am compelled to the belief that we have a different and older group, possibly of Bunter Sandstone Age, in the Spiti Valley.

It will crown all these researches if the Indian geologists should find the shallow and deep water deposits of the Permian Age. It cannot but be that there is a rich harvest, yet unreaped, from this transition period.

---

* Von Buch, Berlin. Trans. 1846, p. 74, plate 5, fig. 2.

JURASSIC ROCKS.

CEPHALOPODA.

By Professor H. F. Blanford,
of Calcutta.

The Jurassic Cephalopoda, described in the following pages, consist of 19 species: one of which is a Belemnite, the remainder Ammonites. The single species of the former genus is certainly identical with that previously described from Cutch and the Oolites of the Spiti Valley, and the differences which distinguish it from the *B. sulcatus* of Miller are so slight that I cannot at present regard it as specifically distinct.

Of the Ammonites, two species only are common to the Oolites of Cutch, viz., *A. biplex*, Sow., and *A. torquatus*, Sow., but the same two species, together with five others, or nearly two-fifths of the entire number, exist in the Spiti collection of Dr. Gerard. Of the Liassic European Ammonites figured by Mr. Everest, and on his authority described by myself as forming part of Gerard's collection, not one occurs in Colonel Strachey's cabinet, and I have now strong reasons for believing that their intermixture with the Spiti fossils was accidental, and that the specimens in question were really from the alum shale of Yorkshire. Except *A. biplex*, Sow., and *A. triplicatus*, Sow., no well-identified European species occurs among Colonel Strachey's fossils, but many of the latter are closely allied to European types, and of these almost exclusively to Ammonites of the Oxford clay. Some of the species, indeed, *A. jubar*, and *A. Griffithii*, Strachey, for example, are but doubtfully separable from European forms.

The majority of the species belong to the section "*Ornati*" of Von Buch, but there are representatives of most of the other Oxfordian types, as will be seen by the following classified list:—

Fimbriati .	*A. alatus.* Strachey.	Armati . .	*A. jubar.* Strachey.
Macrocephali	*A. Nepalensis.* Gray.	Ornati . .	*A. octagonus.* *Id.*
Coronarii .	*A. umbo.* Strachey.		*A. Hookeri.* *Id.*
	A. guttatus. *Id.*		*A. Medea.* *Id.*
Planulati .	*A. biplex.* Sow.		*A. Wallichii.* *Id.*
	A. triplicatus. Sow.		*A. robustus.* *Id.*
	A. torquatus. Sow.		*A. Griffithii.* *Id.*
	A. scriptus. Strachey.	Amalthei .	*A. strigilis.* Blanford.
	A. tenuistriatus. Gray.	. ? . .	*A. acucinctus.* Strachey.

In addition to the measurements of individual specimens, I have, on the suggestion of Colonel Strachey, appended to the description of each species, the relative dimensions of the parts of the shell. In so doing, I do not wish it to be inferred that the ratios given are invariable for each species; indeed, I am convinced that as regards the relative thickness, diameter of the whorls, &c., considerable variation exists in most cases; but comparative measurements, such as those given, afford more useful data for the determination of species, than the mere absolute dimensions of the specimens described. The ratios selected are :—

1st.—That of the radius vector of the ultimate and penultimate whorls, or of the two radii measured from the centre of the spiral to the external edge of the aperture, and to the termination of the second whorl respectively.

2nd.—That of the portion of an inner whorl concealed by the overlap of an outer whorl, *to* the diameter of the latter.

3rd.—That of the thickness of a whorl *to* its diameter, the latter being in all cases understood to be the measurement taken on the plane of the spiral.

P. S.—[in reprint]. Since these pages were written, Professor Dr. Oppel, of Munich has favored me with a copy of his description of the Ammonites collected at Spitih by the late Herr Adolp von Schlagintweit. Some of these species are identical with those here described, and my own nomenclature must therefore yield to that of Dr. Oppel, whose work enjoys priority of publication. Others of Dr. Oppel's species appear to be identical with those of Dr. Gerard's collection, described by myself in the Journ. As. Soc. Bengal, 1863, p. 124. All these synonyms will be found enumerated in a note at the end of this reprint.

Some corrected references to my paper in the Journ. As. Soc. have been inserted in this reprint.

BELEMNITES SULCATUS.—*Miller.*

Plate 10.

Belemnites sulcatus, Miller, Trans. Geol. Soc., 2nd Ser., vol. ii., pl. 8, f. 3.
" *canaliculatus*, Schloth.
" " Sow., Trans. Geol. Soc.
" *sulcatus* Blanford, Spiti fossils, Journ. As. Soc. Beng.

The guards of this species, of which the collection contains a large number of various sizes, are identical with those from the Spiti Valley and Cutch. The sulcation is broader and deeper than in European specimens, and is continued to the apex. With slight variations, the Indian specimens are generally less depressed than those figured by M. D'Orbigny (Pal. Francaise), Zieten, &c., the antero-posterior diameter being nearly equal to the transverse diameter. The largest specimen in Colonel Strachey's collection measures as follows :—

Length	3·6 in.
Antero-posterior diameter	0·9 "
Transverse	0·9 "

Some specimens of phragmocones found with the above, probably belong to the same species. Some of them are figured Pl. 10, figs. 6, 7.

AMMONITES ALATUS.—*Strachey, MSS.*

Plate 18, figs. 3 *a*, *b*.

A. testâ convolutâ, discoideâ, striatâ, lamellosâ; anfractibus rotundatis, depressis, haud celatis, lentè accrescentibus, transversim striatis, intervallìm lamellatis: lamellis rectis, ad basin undulatis, apud umbilicum antrorsum curvatis. Aperturâ transversè ellipticâ.

Diameter of fragmentary whorl	1·2 in.	Spiral ratio	2 : 1
Thickness	1·4 "	Immersion of whorls ...	$\frac{1}{20}$
		Thickness : diameter ...	$\frac{7}{6}$

This is an Ammonite of the "Fimbriati" group, having rounded depressed whorls, increasing slowly in size, and scarcely overlapping. The whorls are transversely striated, and bear at short distances thin prominent lamellæ, frilled at their base, and smooth above, curving forwards towards the umbilicus. In style of ornament, the species closely resembles *A. Rudesianus*, D'Orb., but is of much less rapid growth—a character in which it approaches to *A. Adelæ*, of the same author. The whorls are more depressed than in either of these species, the form of the aperture being transversely elliptical. *A. alatus* is, however, more nearly allied to the Oxfordian than to the Liassic members of the group. As represented in the figure, Plate 18, fig. 3 *a*, which is a restoration compiled from two fragmentary external casts, it appears like a Crioceras; but this appearance is deceptive, and is due to the imperfection of the casts, as a fragment of a whorl, fig. 3 *b*, shows that each whorl is slightly indented by the preceding.

AMMONITES NEPALENSIS.—*Gray.*

Plate 14, figs. 1 *a–b*.

Ammonites Nepalensis, Gray. Hardwicke's Illustr., vol. i., pl. 100, figs. 1, 2.

A. testâ discoideâ, compressâ, costatâ. Anfract. rotundatis, latis, extus compressis, ventre rotundato,—lateraliter complanatis. Costis subflexuosis circa 35, validis, per medium anfractûs plerumque 3-furcatis, apud ventrem prorsùm arcuatis. Umbilico angusto. Aperturâ ovatâ, anticè compressâ.

Diameter	4·8 in.	Spiral ratio...	2 : 1
Ditto of outer whorl ...	2·2 „	Immersion of whorls ...	$\frac{1}{3}$
Thickness	1·9 „	Thickness : diameter ...	$\frac{10}{11}$

This Ammonite is nearly allied to *A. Opis*, Sow., which it resembles in the width of its whorls, and the number and trifurcation of its ribs. *A. Nepalensis* is, however, more tumid, and has broader whorls, and consequently a narrower umbilicus than the Cutch species. The overlap of the whorl is considerably greater than is represented in the figures, Plate 14, and the aperture or section of the whorl, longer and more compressed-ovate than in fig. 1 *b*. The sutures are not visible in either of the specimens examined.

AMMONITES TENUISTRIATUS.—*Gray.*

[Plate 14, fig. 2?] Pl. 15, fig. 2.

A. *tenuisulcatus*, Gray. Hardwicke's Illustrations, vol. i., pl. 100, fig. 4.

A. testâ discoideâ, compressâ, umbilicatâ, crebricostatâ. Anfractibus latis, complanatis; ventre rotundato. Costis flexuosis, per medium anfractûs bifurcatis, apud ventrem antrorsum acutè arcuatis. Aperturâ oblongâ.

Diameter	3·35 in.	Spiral ratio...	2 : 1
Ditto of outer whorl ...	1·4 „	Immersion of whorls ...	$\frac{1}{5}$
Thickness	0·9 „	Thickness : diameter ...	$\frac{2}{5}$

Nearly allied to *A. Nepalensis*, Gray, of which, indeed, I am inclined to think it is only a variety. It is more compressed and flattened than that species, has a somewhat larger umbilicus, and more numerous and finer ribs (about twice as many in the two specimens examined, one of which is Hardwicke's type). The curvature of the ribs is similar in the two species, except that the ventral arcuation is rather more acute in *A. tenuistriatus*. The sutures are not visible.

AMMONITES UMBO.—*Strachey, MSS.*

Plate 17, fig. 2.

A. testâ discoideâ, latè umbilicatâ. Anfractibus depressis, rotundatis, costatis, per medium serie unica tuberculis permagnis ornatis. Costis simplicibus, rotundatis,—nonnullis apud umbilicum, aliis apud tuberculos orientibus: in margine sub-interruptis. Aperturâ reniformi.

Diameter of fragmentary whorl	1·7 in.	Spiral ratio	2 : 1
Thickness	2·1 „	Immersion of whorls ...	$\frac{1}{10}$
		Thickness : diameter ...	$\frac{7}{6}$

An Ammonite of the *A. anceps* and *A. coronatus* type, ornamented with a series of large bossy tubercles on the middle of the whorl, and well-developed, rounded ribs, which are slightly interrupted over the siphuncle. The sutures bear much general resemblance to those of *A. anceps*, but differ in the details, having an additional auxiliary lobe, and a narrower ventral, and broader lateral saddles.

The only specimen in the collection is a fragment of the whorl represented two-thirds of the real size.

AMMONITES GUTTATUS.—*Strachey, MSS.*

Plate 13, fig. 2.

A. *guttatus*, Blanford, Spiti fossils in Journ. As. Soc., Bengal, 1863.

A. testâ discoideâ, latè umbilicatâ, tuberculato-costatâ. Anfractibus sub-depressis, ad marginem latè rotundatis, costatis, apud umbilicum tuberculis 19 compressis coronatis, intervallim 3–4 sulcis profundis notatis. Aperturâ semiorbiculari.

Diameter, circa 3 in.

The above description of this Ammonite is taken, in part only, from the specimen in Colonel Strachey's collection, which is an imperfect external cast of one side of a shell. There can, however, be little doubt that the species is identical with one found at Spiti by Dr. Gerard, and figured in the Journal of the Asiatic Society of Bengal. The latter is the authority for the description of the general form of the shell. The restoration herewith given at Plate 13, fig. 2, is consequently erroneous, the diameter of the whorls being probably at least half as much again as they are represented, while from each tubercle springs a bundle of 4 or 5 ribs, which cross the ventral region with a slight convex curve towards the mouth. Sufficient is seen of a portion of the impression of the outer whorl of Colonel Strachey's specimen to warrant these inferences. *A. guttatus* is nearly allied to *A. coronatus*, Brug, but has a wider umbilicus.

AMMONITES BIPLEX.—*Sowerby.*

Plate 11, fig. 1, Plate 12, fig. 1.

A. biplex, Sow., Min. Conch., Plate 293, f. 1, 2.

„ „ Blanf., Spiti fossils, in Journ. As. Soc., Bengal.

Two specimens of this Ammonite occur in the collection, together with some impressions of the shell on black siliceous nodules. I can detect no difference between them and the characteristic Oxford clay specimens of Europe. They are identical also in all respects (mineral character included) with those from Spiti, lately described by myself, from Dr. Gerard's collection.

AMMONITES TRIPLICATUS.—*Sowerby.*

Plate 13, figs. 1 *a-c.*

A. triplicatus, Sow., Min. Conch., Plate 292 and 293, f. 3, 4.

This Ammonite is only distinguished from the preceding by the fasciculate character of the ribs in adult specimens, young shells of the two species being undistinguishable. It is, however, the case with many allied species of Ammonites, as has been pointed out by M. D'Orbigny, that the specific characters of ornament are only assumed by the adult shell, and I follow, therefore, the majority of English authors in treating provisionally the two forms (with bifid and fasciculate ribs respectively) as distinct.

AMMONITES TORQUATUS.—*Sowerby.*

A. torquatus, Sow., Trans. Geol. Soc., 2nd Ser., vol. v., pl. 21.
A. torquatus Blanf. Spiti fossils, in Journ. As. Soc., Bengal.

The only characters by which I can distinguish this species from *A. biplex*, Sow., are :—Its thicker and more depressed whorls, and a slight notching of the ribs above the siphuncle. These characters are exhibited by the typical Cutch specimens, as well as by those in Colonel Strachey's cabinet,* and also by the specimens described and figured by me, in the Spiti collection of Dr. Gerard. The distinctness of the notching and the depression of the whorls vary, however, in different specimens, and a more extensive comparison is

---

* All the figured specimens of Colonel Strachey's collection have been liberally presented by that gentleman to the Museum of Practical Geology, London.—J. W. S.

requisite to decide whether *A. torquatus* be really distinct from *A. biplex.*

Mr. Sowerby, in his description of the figured specimens from Cutch, states that they are distinct from "a Himalayan species," in having an "incurved inner margin." On comparison of the specimens, I can, however, detect no such difference, nor, indeed, any other than that the Himalayan specimens have uniformly more numerous (about 55) ribs than those from Cutch, which have about 45.

AMMONITES SCRIPTUS.—*Strachey, MSS.*

Plate 16, figs. 2 *a-c.*

A. testâ discoideâ, compressâ, tuberculatâ, costatâ. Anfractibus rotundatis, vix complanatis, circâ umbilicum tuberculis mammæformibus coronatis, è quibus costæ binæ oriunt. Costis valdè depressis, antrorsum arcuatis, apud medium anfractûs bifurcatis, ad marginem obsoletis. Umbilico magno. Aperturâ semi-ellipticâ.

Diameter of fragmentary whorl	1·5 in.	Spiral ratio	5 : 3
Thickness	1·3 „	Immersion of whorls ...	$\frac{3}{7}$
		Thickness	$\frac{6}{7}$

A discoid form, with whorls nearly as thick as broad, rounded ventrally, and somewhat flattened at the sides; ornamented along the umbilicus with a row of about 20 depressed rounded tubercles, from which spring pairs of ribs, also but little developed, and which curve forward across the whorls, bifurcating about half-way between the tubercles and the periphery, where they become nearly obsolete. The sutures (fig. 2 *c*) are 5-lobed on each side, the ventral lobe as broad as, or broader than the adjoining saddles. These saddles are rounded, and symmetrically divided. The superior lateral lobe is sub-symmetrical and trifurcate.

A. scriptus is nearly allied to *A. guttatus* but is distinguished by its more compressed form, as well as by the regular bifurcation of its ribs, and some differences in the details of the sutures. The only specimen in the collection is the fragment figured.

AMMONITES JUBAR.—*Strachey, MSS.*

Plate 20, figs. 2 *a-c*. Plate 21 figs. 1 *a-c*.

A. testâ discoideâ, compressâ. Anfractibus complanatis, costatis, intervallim sulcatis, et apud peripheriam tuberculis binis depressis notatis. Costis flexuosis, bifidis vel 3-4 furcatis, ad marginem sub-obsoletis. Ventre planato. Aperturâ truncata-ellipticâ. Septis lobis lateralibus 4.

Diameter . . .	3·3 in. (approx.)	Spiral ratio . . .	. 5 : 2
Ditto of outer whorl .	1·3 „ „	Immersion of whorls .	. $\frac{1}{6}$
Thickness . .	0·9 „ „	Thickness : diameter .	. $\frac{2}{3}$

This is a closely ribbed flat-whorled Ammonite, bearing a very near resemblance to *A. Bakeriæ*, Sow., from which it is distinguished only by the greater width and flatness of its whorls, and by the bending forward of the ribs on each side of the siphuncle. The ribs are filiform and flexuous, varying in number from 45 to 55, mostly bifid, and in old specimens irregularly fasciculate. They bend forward considerably in the ventral region (in some specimens more than in others), and become obsolete on the outer or median line. The stages of growth are marked sometimes by sulci and strong oblique varices, and in all cases the ventral region bears several pairs of low flat tubercles, (one on each side,) which occur at regular distances, the intervals varying in different specimens. The sutures resemble those of *A. Parkinsoni*, Sow., and *A. Wallichii*, Gray, but the ventral lobe is, relatively to the superior lateral, shorter than in the last-named species, while the superior lateral differs from that of both species in being symmetrically trifid.

Var. *a multiradiatus*. Strachey, MSS.

This differs from the normal form in the greater number of its ribs (55), which are consequently more close set and filiform. The variation probably occurs only in young shells.

AMMONITES OCTAGONUS.—*Strachey, MSS.*

Plate 12, figs. 2 *a-b.*

A. octagonus, Blanf. Journ. As. Soc., Bengal.

A. testâ discoideâ, compressâ. Anfr. paullo amplexantibus, subcompressis, ad latera seriebus 3 tuberculorum magnorum ornatis, inæquicostatis. Aperturâ ellipticâ. Septis lateraliter 3-lobatis

Diameter of fragmentary whorl	1·3 in.	Spiral ratio	5 : 3
Thickness	1·1 „	Immersion	$\frac{1}{5}$
		Thickness : diameter	5

The only specimen of this Ammonite in Colonel Strachey's Collection is a fragment, but of larger dimensions, and in better preservation than that previously described from Spiti; and it exhibits characters which lead me to relinquish my former idea of its identity with the *A. Eugenii* of Raspail. The fragment in question is a portion of the body whorl (the inner whorls being wanting), and is ornamented with three rows of very prominent tubercles on each side. The tubercles of the two inner series are connected in pairs, by coarse ribs, inclined somewhat forwards across the whorls. Those of the outer series are connected in like manner with those of the middle series, but are more numerous, so that two of the former are sometimes connected with one of the middle ones, a character not noticed in *A. Eugenii.* Some of the outer ribs show a tendency to a chain-like duplication; but this feature is somewhat exaggerated in the figure.

AMMONITES HOOKERI.—*Strachey, MSS.*

Plate 17, figs 1 *a-d.*

A. testâ discoideâ, umbilicatâ, tuberculato-costatâ; anfractibus octagonis. Costis validis, anticè arcuatis, apud peripheriam abrupt-interruptis; tuberculis—apud umbilicum lateraque nonnullis mage nis; apud peripheriam omnibus compressis; in serie umbilicali prominentibus. Costis â tuberculis lateralibus bifurcatis. Umbilico aperto.

Diameter of whorl, 1 inch. Thickness of ditto, 1 inch.

An openly-umbilicated shell, ornamented with strong ribs, most of which spring from a series of large tubercles, about twelve or fourteen in number, near the umbilicus, and bifurcate at another series about half-way across the whorl, equal in number to those along the umbilicus. A few non-tuberculate and simple ribs intervene, and all terminate on each side of the siphuncle in a row of compressed tubercles. The superior lateral lobe is longer than the ventral;—broad, and terminating in three branches, of which the central is the longest.

The species bears much resemblance to *A. octagonus*, Strachey, but has not the catenate ribs of that species, while the ventral tubercles are much more numerous.

AMMONITES MEDEA.—*Strachey, MSS.*

Plate 19, fig. 5 *a*.

The only specimen of this MSS. species of Colonel Strachey in his collection is the fragment figured, from which it is difficult to pronounce upon its affinities. It may be either, as surmised by Colonel Strachey, a species allied to *A. Jason*, Zieten, or a portion of a large specimen of the tuberculate form of *A. Wallichii*, Gray.

AMMONITES WALLICHII.—*Gray.*

Plate 15, figs. 1 *a-c*; Plate 19, figs. 1 *a-c*, 2 *a-c*.

Ammonites Wallichii, Gray, Hardwicke's Illustr., vol. i., plate 100, fig 3.
" " Blanford, Journ. As. Soc. Beng., 1863.

A. testâ discoideâ, compressâ, flexuoso-costatâ. Anfr. plus minusve planatis, medio ventre excavatis. Costis angustis plerumque bifurcatis apud ventrem interruptis, nonnunquam tuberculatis (senioribus nonnullis continuis). Umbilico lato. Aperturâ ovatâ anticè truncatâ. Lobis lateralibus septorum 4.

Diameter	* 3·75 in.	
Ditto of outer whorl	1·35 "	
Thickness	1·20 "	
Spiral ratio	2 : 1	to 5 : 2
Immersion of whorls	$\frac{1}{7}$	" $\frac{1}{6}$
Thickness : diameter	$\frac{7}{8}$	" $\frac{3}{4}$

* A specimen in the British Museum measures not less than six inches in diameter.

The ribs of this species terminate abruptly on each side of the siphonal region, leaving a smooth excavated space between. In very old specimens, they are, however, continuous across this area. The ribs are, for the most part, either simple or bifid, but, in some specimens, two or more spring together from the umbilical margin. The proportions of the whorls are somewhat variable, the breadth of the outer whorl ranging from $\frac{3}{100}$ to $\frac{40}{100}$ of the diameter, and the thickness from $\frac{28}{100}$ to $\frac{34}{100}$. The ventral lobe of the sutures is rather shorter than the superior lateral, which is broad, open, and subsymmetrical.

A variety of this species is figured at plate 19, figs. 1 *a–b*, the ribs of which terminate ventrally in two rows of tubercles. The shell is much compressed, and the ribs numerous and regularly bifid.

AMMONITES ROBUSTUS.—*Strachey, MSS.*

Plate 16, figs. 1 *a-c*.

A testâ plus minusve globosâ, valde costatâ, senectute tuberculatâ. Anfractibus amplectentibus, add umbilicum crassioribus, apud ventrem compressiusculis, rotundatis: lateribus subplanatis. Costis validis, irregularibus, rursum ac rursum bifurcatis, apud $\frac{3}{4}$ diametri anfractûs intervallim tuberculatis in medio ventre interruptis. Aperturâ semicirculari, subtrigonâ, vel. sagittatâ. Umbilico minimo. Lobis lateralibus 5.

	a.	b.	c.
Diameter	3·1 in.	2·2 in.	2·4 in.
Ditto of outer whorls ...	1·8 „	1·15 „	
Thickness	1·6 „	1·5 „	0·8 „
Spiral ratio		20 : 9	
Immersion of whorls		$\frac{1}{5}$	
Thickness : diameter		$\frac{7}{8}$ to $\frac{3}{4}$	

A very variable form, as may be inferred from a comparison of the above measurements. It is sometimes much compressed, with a narrow, almost keeled ventral region;—sometimes globose. The ornament is characteristic. The ribs bifurcate repeatedly at various parts of the whorl, and are interrupted in the middle of the ventral region, leaving a narrow sulcus.

In the adult, every third or fourth rib is furnished with a prominent tubercle, at about three-quarters the diameter of the whorl, where the ribs finally bifurcate.

In the figures, two different specimens are represented. Fig. 1 *b* is tolerably correct, except that the ventral sulcus is exaggerated below. In fig. 1 *a*, the tubercles should be placed only at every third or fourth rib, and somewhat nearer to the umbilicus, and a bifurcation should be represented at every tubercle.

AMMONITES GRIFFITHII.—*Strachey, MSS.*

Plate 20, figs. 1 *a–c*.

A. testâ discoideâ, compressâ, complanatâ, latè umbilicatâ, costatâ, apud ventrem seriebus 2 tuberculorum ornatâ, inter quæ venter excavatus. Costis numerosis, irregulariter bifurcatis, apud ventrem continuis. Aperturâ ovato-oblongâ. Septis lateraliter 3-lobatis. Lobo ventrali brevissimo.

Diameter	4·7 in.	Spiral ratio	2:1
Ditto of outer whorl ...	1·7 „	Immersion of whorls	$\frac{1}{5}$
Thickness	1·1 „	Thickness : diameter	$\frac{5}{8}$

As represented in the figure, the two rows of tubercles which ornament the ventral region of this shell are seen covered up by the adhering shell of the succeeding whorl, and thus appear to be two keels, with a smooth intervening excavated area. Another part of the shell explains the real state of the case, viz., that the ribs are continuous across the siphuncular region, and bear two rows of smaller tubercles, similar to those of *A. Duncani*, D'Orb., to which the present species is nearly allied, if indeed the two forms be not mere varieties of the same species. The ventral lobe is not more than half as long as the superior lateral, which last is as broad as the ventral saddle and terminally trifurcate.

Only one specimen, viz., that represented in the figure, occurs in the Collection. It is an imperfect shell, and bears the remains of three-fourths of another whorl. The figure is three-fourths of the real size of the specimen.

AMMONITES STRIGILIS.—*Blanf.*

Journ. As. Soc. Beng., 1863.

A single specimen (cut in half and polished) of this Ammonite (without specified locality) only differs from the original Spiti specimen in its somewhat larger size.

Diameter of shell, 2¼ in. Diameter of outer whorl, 1¼ in.

AMMONITES ACUCINTUS.—*Strachey, MSS.*

Plate 18, figs. 1 *a–c*; plate 19, fig. 4 (var. plate 18, fig. 2).

A. testâ discoideâ, percompressâ, complanatâ, levigatâ, vel striis rugisque falciformibus obsoletè notatâ. Ventre sub-acuto, ad medium minutissimè denticulato. Umbilico formæ normalis angustissimo. Aperturâ obtusè sagittatâ.

Diameter	1·5 in.	Spiral ratio...	5 : 2
Ditto of outer whorl ...	0·9 „	Immersion of whorls ...	¼
Thickness	0·4 „	Thickness : diameter ...	⅖ (approx.)

A compressed discoid Ammonite, with flat sides and a very small umbilicus somewhat variable in diameter, but averaging in type specimens about one-tenth of that of the shell. The sides are ornamented with very numerous, regular falciform ribs, tolerably distinct in some specimens, nearly obsolete in others, and with a row of minute teeth forming an interrupted ventral ridge over the siphuncle. The ventral lobe is about two-thirds as long as the superior lateral, and rather broad. The other lobes sub-symmetrical and terminating in three digitate branches. The saddles are symmetrically bifid.

The figure of this species given in plate 18 is, in some respects, erroneous. The ribs in figs. 1 *a* should conform to the shape of the mouth, instead of being but slightly flexuous, and the periphery should be ornamented with minute sharp teeth, instead of crenulations.

Var. *a*. Plate 18, figs. 2 *a-b*.

A. mundus. Strachey, MSS.

This differs from the normal form in its wider umbilicus, the diameter of which amounts to one-fourth that of the shell. The two fragments in the collection are those of larger shells than any of the normal form, and although ornamented with the denticulate ridge on the inner whorls, the outer whorls are smooth and rounded on the ventral surface.

[Since this was in type, the figures have been corrected (as far as the state of the lithographic stones would allow), in conformity with Prof. Blanford's instructions.—J. W. S.]

JURASSIC GASTEROPODA AND BIVALVES.

By J. W. Salter, F. G. S., &c.

The description of these few species has fallen to my lot, only because my friend and fellow-labourer, Professor Blanford,* has left England for his important duties at Calcutta. He sent me the last MSS. of the Ammonites, &c., &c., by steamer, to Southampton, and left me to finish the rest. All the Oolitic fossils are from the same range of strata; but, as he has shown in his late Paper in the Asiatic Society's Journal, these strata are very thick, and probably include several members of the Jurassic series. In fact, if we did not know this from the published sections, the fossils would indicate it. They certainly include some forms of the Lower and Middle Lias, and of the Inferior and Great Oolite; and his identifications include the Oxford Clay.

The two Gasteropods are hardly worth a separate notice. One seems new. I call them Turbo and Chemnitzia. Professor Blanford finds Gasteropoda equally scarce *(Pleurotomaria, Turritella)* in the Spiti Pass.

TURBO INVITUS.—*N. Sp.*

Plate 21, fig. 3.

This minute and solitary specimen probably belongs to the genus Turbo, but its characters are very obscure. It is not four-tenths of an inch high; the diameter about equal to the height; the whorls four, rather convex, and showing three spiral ridges—the upper near the rather deep suture, the

* Of the Bengal Club, Calcutta.

second and third about equi-distant from each other and the first, and placed on the most prominent part of the whorl,—one not projecting beyond the other. Base convex; lines of growth?

The Litorinæ and Turbo in our Oolites have seldom so few ribs, and they are generally more roughly ornamented.

CHEMNITZIA.—*N. Sp.*

Plate 21, fig. 2.

A minute fossil, scarcely one-third of an inch long, and very like numerous other species in Oolitic rocks. The whorls are simply rounded, not appressed at the suture, nor oblique. The mouth is ovate and rather pointed below.

The specimen has been lost, and only a general resemblance can be traced to such species as *C. lineata*, Sowerby, which appears smooth in its younger state, and is found both in Oolite and Lias. Such species are, however, found in all rocks. From the Lower Lias the Abbé Stoppani has figured a small Chemnitzia and a Pleurotomaria, which remind us of the two above species.—Palæont. Lombardie, 3rd Series, pl. 2.

OSTREA FLABELLOIDES.—*Deshayes.*

Plate 22, fig. 1.

Goldfuss., Pet., plate 73, figs. *f–i.*

Our fossils are but small, and internal casts only. I should have been rather disposed to refer them to the *O. Marshii*, Sow., a fossil already described by Sowerby himself from the Cutch district. But the ribs of *O. Marshii* are so very sharp and high, and made so nodular by the ridges of growth, that I prefer identifying with the Inferior Oolite species. *O. flabelloides* has the shape longer, and somewhat rounder and lower ribs, and between these ribs are curious pit-like hollows, giving a somewhat bladdery appearance. I think it may be safely identified.

O. gregaria, from the Kelloways Rock and Coral Rag, is too finely ribbed; else it is a good deal like our shell in general character. *O. nodosa*, Goldf., has too many ribs.

O. ACUMINATA.—*Sowerby.*

Plate 22, fig. 3.

Mineral Conchology, plate 135, fig. 2.

Exactly like the British fossils from Cornbrash, Great Oolite, Inf. Oolite, &c. The Bradford clay species, *O. Sowerbyi*, is a good deal like, but far too broad and too little curved.

AVICULA INŒQUIVALVIS.—*Sow.*

Plate 22, fig. 13.

Mineral Conchology, plate 244.

Though the ribs are a little more spinulose, I still think these specimens may be identified. The chief point of difference is that the ribs are somewhat rougher, and that there are no intermediate ones. In this respect it is like *A. decussata*, a species from the Lower Lias, but which has more numerous ribs. *A. echinata*, Sow., is too oblique a species to compare with ours.

MONOTIS CONCENTRICUS.—*Blanford.*

Plate 22, fig. 4.

Asiatic Soc. Journal, 1862, plate 4, figs. 6, 7.

This remarkable shell reminds us at first sight of the small Inocerami of the Gault formation, and again, remotely, of the so-called *Inoceramus dubius*, Sow., of the Lias.

But Mr. Blanford has detected, in specimens from the Spiti Pass, the characteristic sinus and anterior ear, and we have endeavoured to insert it in the figure (4 *a*).

I quote Professor Blanford's paper, published last year; the plates are perhaps not yet published, as they were finished this year in London. He says, "In form and sculpturing of the left valve it much resembles *Inoc. concentricus* of the Gault. The right valve is like an *Anomia,* with a narrow linear notch just under the hinge line."

Our specimen is finer than the one figured in the journal above quoted. It will be seen, by comparing Mr. Blanford's list with ours, how many of the Spiti fossils are identical with those detected by Strachey in the Niti Pass. (*See* pp. 93, 94.)

PECTEN COMATUS.—*Munst.*

Plate 22, fig. 9.

Goldfuss, Petref., t. 91, f. 4.

This agrees almost as well with *P. obscurus,* Sow., as with the above, and I hardly know to which to refer it. The shape is more like *P. comatus,* and the ears not so oblique as in Goldfuss' figure of the other Lias species.

P. arcuatus, Sow., of the Forest Marble of England, is like our shell, but is less convex, more oblique, more oval, and has the striæ much coarser. Indeed, there are several Oolitic forms quite like ours. *P. Lens,* Sow., is rather too round a shell.

PECTEN BIFRONS.—*N. Sp.*

Plate 22, figs. 5—7.

Half an inch long, and rather less broad: ovate, excluding the broad ears, which are four lines broad. The posterior ear is sunk, and has strong ridges of growth, but very faint radiating lines. The shell is gently convex.

The ribbing is peculiar, and wholly different on the internal and exterior aspect of the shell. Within (fig. 5 *c*) there are about nine strong ribs most prominent at the margin. Outside, the more convex valve has at least twice as many, each rib being divided into two strong ridges, and some of these are again duplicated and occasionally interlined (figs. 6, 7).

The flatter valve (figs. 5, 5 *b*) has much more numerous ribs, generally three to an internal rib, and these are much more equal in thickness than in the other valve. All are crossed by lines and ridges of growth, and well preserved specimens show distinct but slender striæ crossing the ribs in a regular manner, and decussating them. Fig. 7 *b* does not show this character so perfectly as would be desirable.

This pretty species will agree with none in the British Collections; but *Pecten paradoxus*, (Goldfuss, plate 99, fig. 4), from the Lias of Bavaria, is like it, and *P. Massalonghi*, figured in the Abbé Stoppani's work on the Lower Lias, or Kossen beds of Lake Como, has much the same character. Neither have the coarse *external* ridges.

PECTEN MONILIFER.—*N. Sp.*

Plate 22, fig. 10.

A species not remotely allied to such species as *P. articulatus*, Schloth., and less closely to such as *P. barbatus*, Goldf. From the former, an inferior Oolite species, it is at once distinguished by the few ribs, not above half the number occurring in the Himalayan form. Nor is the ornament quite like.

We have three specimens, all imperfect. The shell was more than half an inch long, and about as broad, with a generally oval appearance. It is convex in the larger valve, and nearly flat in the other. Both valves have five or six large angular ribs, the interstices being angular likewise, and crossed by very fine, close lines of growth. The side ribs become duplicated, having smaller ribs of like character on their flanks (fig. 10 *b*). The lines of growth are stronger and more rugged in some specimens than others. The main ribs are strongly angular, and ornamented by equi-distant tubercles or short spines, so regularly placed as to suggest the name.

We know nothing of the ears, beak, or interior. But the ornament is peculiar; and as my friend Mr. Etheridge knows no species but *P. barbatus* that can be compared with it, I have thought it should receive a name.

P. SABAL.—*N. Sp.*

Plate 22, fig. 11.

A remarkable species, for which we find no good analogue in our Oolitic series. The ribs are rounded, and truly triplicate, each stronger rib being flanked by two minor ones: the interstices are filled by longitudinal, very fine striæ, and crossed by close sharp lines of growth. One ear is sunk, and has the concentric lines. We do not know the other at all. Nor, if fossils were more plentiful from eternal snows, would the species have been named here at all.

P. LENS.—*Sow.*

Plate 22, fig. 8.

Mineral Conch., plate 205, figs. 2, 3.

The apical angle of the beak and the general form agree well with Sowerby's species from the inferior Oolite, and the surface markings are quite like, the minute squamose lines of growth, however, not being mixed with larger ones, as in the British fossil. I think it is merely a question of preservation.

P. cingulatus, Sow., from the Marlstone, is like, but rounder, with a larger apical angle (more than a right angle—ours is less) and the surface polished.

PECTEN.—*Sp.*

Plate 22, fig. 12.

Not perfect enough to identify, or worth figuring, did it not occur on a slab with *Cardium truncatum*, Sow.

LIMA GIGANTEA.—*Sow ?*

Mineral Conch., plate 77.

I cannot separate one fragment from the Niti Pass, from Sowerby's well-known Lias fossil.

LIMA ACUTA.—*Stoppani ?*

Plate 22, fig. 14.

Palæontologie Lombardie, 3rd ser., plate 13, fig. 9; or compare with *Lima punctata*, Sow. Min. Conch., plate 113, figs. 1, 2.

I think this agrees best with Stoppani's figure; but unless the radiating lines are rather too strong, it agrees as well with the common Marlstone fossil, *L. punctata*. Mr. Etheridge would identify it with this latter species.

INOCERAMUS HOOKERI.—*N. Sp.*

Plate 23, fig. 1; and Var. CRENATULINUS, fig. 2.

A fine shell, nearly three inches long, and more resembling *I. concentricus* of the Gault, or *Inoc. lingua*, Goldfuss, from the Greensand, than any other species.

We have two forms,—possibly two species. Fig. 2 appears to be remarkably narrower than fig. 1, and to have a different curvature of the ribs. I have, therefore, distinguished it, but at present only as a variety.

The broader form, fig. 1, has a small posterior ear on which the lines of growth are sharp, but not ridgy. The surface is occupied by broad prominent ribs, with equally broad furrows between them, and the ribs are of the same thickness and strength from side to side of the shell. The lines of growth do not coincide with the ribs.

The narrower form, fig. 2, has a pointed beak, a short oblique hinge-line, a less regularly convex surface abruptly and steeply bent down along the anterior side, on which side the concentric ridges are lost—lines of growth only being present at intervals.

The ridges are very prominent, but narrower than the interstices, sharp-edged above, so as to be somewhat imbricated upwards; and their greatest curvature, quite a sharp bend, takes place along the middle of the disk; a circumstance which does not occur in the larger variety.

Both varieties occur in the same dark cherty nodules.

MODIOLA.—*Sp.*

Plate 23, fig. 3.

I would not attempt to name a *Mytilus*, or *Modiola.* But this is far from unlike a species figured by the Abbé Stoppani in his Palæontologie Lombardi, from the Lower Lias. *M. psilonoti,* Quenstedt, is the species here referred to. It is too broad posteriorly for our fossil, and the anterior side too small. *Modiola anatina,* Smith, a Fuller's earth species, is the nearest British type.

MYOPHORIA BLANFORDI.—*N. Sp.*

Plate 23, fig. 6.

Compare *M. inflata,* Emm. (*Trigonia postera,* Quenst.), Stoppani, Pal. Lomb., 3rd Ser., pl. 7, f. 4, 5; and *Lyrodon curvirostre,* Goldf., pl. 135, f. 13.

Triangular, moderately convex, rather thick as broad as long [Surface—?]. Posterior side much the largest, gently convex, divided by a strong carina (which is produced into a beak at the posterior angle) from the disk. Within this carina the posterior slope is first convex, then concave to the margin : the two areas being separated by a faint ridge. In front of, and close to the carina, the disk has one or two slight ridges. But no trace of outside ornament is left.

The characters above assigned to the posterior area contrast with those of *M. liassica,* Stoppani, l. c. figs. 6-10. Where ours is convex, the Italian species is concave, and *vice versâ.*

But *M. inflata,* Emmr., does appear, from Stoppani's figure, to be very like ours in the cast: except that his description and figure agree in giving little importance to the posterior carina, so conspicuous here; and our species besides shows two ridges behind the carina.

NUCULA.—*Sp.*

Plate 23, fig. 8.

A less convex species decidedly, but very much of the type of *N. variabilis.*, Quenst., from the Lower Lias of Britain.

MACTROMYA or COR   EA?

Plate 23, fig. 7.

Very imperfect and not worth a figure. We have two specimens.

CARDIUM TRUNCATUM.—*Sow.*

Mineral Conch., pl. 553, fig. 3.

Occurs on the same slab as the Pecten above noticed, p. 86. The specimens are imperfect, but still recognizable. Mr. Etheridge identified it directly.

ASTARTE or MYACITES?

Plate 23, fig. 10.

With so many genera to choose out of, from Oolitic rocks, it is impossible to say of what genus this shell may be.

It might be a *Tancredia*, a *Corbicella*, a *Capsa?* or a *Myacites*—or it may even be an *Astarte* of unusual type.

ASTARTE UNILATERALIS.—*Sow.*

Plate 23, fig. 10.

Sowerby in Geol. Tr., 2nd Ser., vol. v., pl. 21, f. 14.

The lateral beaks and regular concentric ridges find their parallel in *A. lurida*, Sow., an Oolite shell which has a wide range. That species, however, shows nothing of the curiously-flattened umbones, a character but ill-expressed in our figure.

CUCULLÆA LEIONOTA.—*N. Sp.*

Plate 23, fig. 4.

Above 2 inches long, and $1\frac{1}{4}$ broad, of a very regular transverse oval shape, only a little more pointed behind than in front, and with but a slight posterior ridge. The whole shell is smooth, even, gently convex, and regular in shape: and for the genus remarkably so.

The beak is placed rather behind the anterior third: it is small and little prominent. The anterior side is regularly convex, and rounded from beak to ventral margin, without depression for a lunette, or any sinus. The front margin regularly rounded. The posterior side slightly more pointed than the anterior, but regularly rounded, nevertheless, into the hinge-margin. A low blunt diagonal ridge marks off the posterior side. Hinge plate moderately broad, with about three long diverging posterior, and three or four anterior, teeth. Pallial line at some distance from the margin.

The surface is marked only with a few lines of growth, antiquated on the posterior slope. Except this, the shell is quite smooth; free even from microscopic lines. And this is so rare in the genus, that there is no occasion to compare it with other species.

ANATINA VAGINULA.—*N. Sp.*

Plate 23, fig. 5.

A fragment of an elongate bivalve, with a strong subcardinal ridge near the anterior end, would have suggested several affinities, either of which might be a true one. But the Abbé Stoppani, in his work, already quoted, has figured an *Anatina* from the Upper Trias, which sets the matter at rest.

A. vaginula differs from this Trias species—(*A. triassica*, Stopp., l. c. pl. 16, f. 17) in being only half the proportional length. Ours was sixteen lines long when a perfect shell, and full half an inch broad at the beak.

The anterior side [is lost]. The posterior end is regularly attenuate, and rather quickly sloping beyond the hinge line on the dorsal margin, which, however, is not much curved: a faint oblique ridge or two occur beneath it, but only on the posterior slope; and there appear to be no internal radiating ridges, as in the Alpine form. *A. precursor*, Stoppani, is a curved species, and does not need close comparison.

TEREBRATULA NUMISMALIS.—*Lam.*

Plate 28, fig. 4.

Lamarck, Anim. sans Vert., 1819, vol. vi., No. 22.
T. orbicularis, Zieten, 1832. Verst., Wurtemb., pl. 39, f. 5.
T. numismalis, Bronn; Von Buch; Quenstedt; D'Orbigny; Davidson, &c.
T. cor. Valenciennes.

Our single specimen agrees well with ordinary specimens from the English Marlstone, except in being a little more elongate, and the beak rather more prominent. It has the characteristic ridges of growth, which are not very well shown in Davidson's figures.

T. CARINATA.—*Lamarck.*

Plate 21, fig. 5.

Compare with *T. resupinata*, Sow.; *T. Moorei*, Davidson; *T. emarginata*. Sow.; and *T. impressa*, Von Buch—in Davidson's Monogr. of Oolitic Brach., Pal. Soc., 1852, pl. 4.

Our artist has omitted the most important character of this shell, viz., the depression of the upper valve upon the front margin. A narrow, but rather deep sinus, hidden from sight on a front view, shows our shell to belong to the group of the *T. resupinata* and *T. carinata*, so common in the Oolites. But I cannot identify it certainly with any, and I should be as sorry to describe a new species of *Terebratula*, as to invent a new tax. I think it is nearer to *T. resupinata* than any other, or perhaps to *T. Moorei*, from its very convex upper valve. But the depressed marginal sinus is unlike either, and more resembles that of *T. emarginata*. *T. carinata* is nearest in external form to our shell, the beak of which, however, is much less carinate on the sides: its upper valve convex and unfurrowed to the very margin, which is not quite the case even in *T. carinata*. But the shape is more rounded, and in this particular it comes nearer the *T. Moorei*, Davidson.

TEREBRATULA GLOBATA—*Sow.*

Plate 21, fig. 6.

Mineral Conchology, 1825, pl. 436, f. 1. *Terebr. Kleinii*, Morris, Deshayes. D'Orb. Bronn, (not of Lamarck). Davidson Monogr. l. c. pl. 13, f. 2-7.

Our specimens have the plait on the front, just a little more pronounced than in British examples, and the back a little more carinate. I can see no differences of importance, and Mr. Etheridge agrees with me.

RHYNCHONELLA CONCINNA.—*Sow.*

Plate 21, fig. 8.

Terebratula, Mineral Conch. 1812, vol. i., t. 88, f. 6. Von Buch, Morris, Bronn, &c. *Rhynchonella*, D'Orbigny; Davidson in Monogr. l. c. t. 17, f. 6-12.

This common species in the Great Oolite and Bradford clay varies much, but is not often quite so large as the Himalayan forms. The species appears, from what I saw in the Oxford Collection, to be equally common in the Spiti Pass.

R. VARIABILIS.—*Schloth.*

Plate 21, fig. 7.

T. variabilis, Schlotheim, 1813; Von Buch; *Rhynchonella* D'Orbigny; Davidson, l. c. t. 16; f. 1-6 and 15, f. 8-10. *T. triplicata* and *T. bidens*, Phill.; Desh.; Deslongch.; Morris; Bronn.

I prefer to consider this, as Mr. Etheridge also does, a variety of the common Lias fossil. In fact, our specimens more resemble Davidson's most characteristic figures than do British specimens from the Lower Lias; the species being more characteristic of the Upper Lias and Marlstone. We have specimens with 2, 3, and 4 raised plaits on the front. Our figure is not quite satisfactory.

Rhynchonella varians of Schlotheim is a different shell, and much more nearly like *R. concinna*.

PENTACRINITES.—*Sp.*

Plate 23, fig. 11.

P. basaitiformis, Miller, a Lias species, is the nearest to our fossil.

TOOTH OF—?

Plate 23, fig. 12.

LIST OF THE HIMALAYAN OOLITIC FOSSILS FROM THE NITI AND SPITI PASSES.

By Professor Blanford and J. W. Salter.

Himalayan species.	Localities in India.	Formations in which the species, or their near allies, are known in Europe. Trias.	Lias.	Inf. Oolite.	Gt. Oolite.	Oxf. Clay.	Localities in Europe.	Nearly-allied species in Europe.
	Niti							
Belemnites sulcatus, Miller ...	Spiti	...	...	×				
	Cutch							
Ammonites acucinctus, Strachey ..	Niti							
	Spiti							
„ var. mundus, Strachey	Niti							
„ alatus, Str. ...	Niti	...	×	...	...	×	...	A. Rudesianus, D'Orb.
								A. Adelæ, id.
„ bifrons, Brug. ...	Spiti	...	×	..	...	...	Europe	
„ biplex, Sowerby ...	Spiti	...	...	...	...	×	Europe	
	Niti							
„ communis, Sow. ...	Spiti	...	×	...	...	...	Europe	
„ concavus, Sow. ...	Spiti	...	...	...	×	...	Europe	
„ Eugenii, Rasp. ? ...	Spiti	...	×	...	...	×	Europe	A. Taylori, Sow.
								A. perarmatus, Sow.
„ Gerardi, Blanf. ...	Spiti	×	×	...	...	...	...	"Globosi" group.
„ Griffithii Str. ...	Niti							
„ guttatus, Str. ...	Niti	...	×	×	...	...	...	A. coronatus, Brug.
	Spiti							
	Cutch							
„ Hookeri, Str. ...	Niti	...	...	...	×	×	...	A. octagonus, Str.
„ heterophyllus, Sow. ...	Spiti	...	×	×	...	...	Europe	
„ Himalayanus, Blanf....	Spiti	...	×	×	×	×	...	"Ceratites" group.
„ Hyphasis, Blanf. ...	Spiti	...	×	×	...	...	...	A. Parkinsoni, Sow.
								A. communis, Sow.
„ Jubar, Strach. ...	Niti	...	...	×	...	...	...	A. Parkinsoni, Sow.
	Spiti							
var. multiradiatus	Niti							
„ Medea, Strach. ...	Niti	...	...	...	...	×	...	A Jason, Rein.
„ Nepalensis, Gray ...	Niti	...	...	...	...	...	...	A. Opis, Sow.
	Spiti							
„ octagonus, Str. ...	Niti	...	...	...	...	×	...	A. Eugenii, Rasp.
	Spiti							
„ robustus, Str. ...	Niti							
„ scriptus, Str. ...	Niti	...	...	...	×	...	...	A. guttatus, Str.
„ Spitiensis, Blanf, ...	Spiti	...	...	...	×	...	...	A. guttatus, Str.
„ strigilis, Str. ...	Niti	×	×	...	...	...	...	"Globosi" or "Amal-
	Spiti							thei" groups.
„ tenuistriatus, Gray ...	Niti							
„ torquatus, Sow. ...	Niti	..	...	...	...	×	...	A. biplex, Sow.
	Spiti							
	Cutch							
„ triplicatus, Sow ...	Niti	...	...	...	...	×	Europe	
„ Thouarsensis, D'Orb.	Spiti	...	×	...	...	...	Europe	
„ (A. Acteon, D'Orb.)								
„ umbo, Str. ...	Niti	...	×	×	...	...	...	A. coronatus, Brong.
„ Wallichii, Str. ...	Niti	...	...	×	...	...	...	A. Parkinsoni, Sow.
	Spiti							
Turritella montium, Blanf. ...	Spiti							

Himalayan Species.		Localities in India.	Formations in which the Species, or their near allies, are known in Europe. Trias.	Lias.	Inf. Oolite.	Gt. Oolite.	Oxf. Clay.	Localities in Europe.	Nearly-allied species in Europe.
Pleurotomaria, 2 species	...	Spiti							
Turbo invitus, Salter	...	Niti							
Chemnitzia—sp.	...	Niti	...	×	×				
Anatina vaginula, Salter	...	Niti	...	×	...	...	...	...	A. triassica, Stopp.
Myophoria Blanfordi, Salter	...	Niti	...	×	...	...	...	...	M. inflata, Emm.
Cardium truncatum, Sow.	...	Niti	..	×	...	...	...	Europe	
Cyprina trigonalis, Blanf.	...	Spiti							
Astarte major, Sow.	...	Cutch	...	...	×				
		Spiti							
„ unilateralis, Sow.	...	Cutch	...	...	×	×	...	...	A. lurida, Sow.
		Spiti							
Modio'a—sp.	...	Niti	...	×	...	...	...	...	M. psilonoti, Quenst.
Nucula cuneiformis, Blanf.	...	Spiti							
		Cutch	...	...	×				
Cucul!æa virgata, Sow.	...	Cutch							
		Spiti	...	...	×				
„ leionota, Salter	...	Niti							
Inoceramus Hookeri, Salter	...	Niti	...	...	×	?			
„ var. crenatulinus,	id.	Niti	...	...	×	?			
Lima acuta, Stopp.	...	Niti	...	×	...	...	...	N. Italy	
„ gigantea, Sow.	...	Niti	..	×	...	...	...	Europe	
„ mytiloidea, Blanf.	...	Spiti	...	×	...	...	...	...	L. gigantea, Sow.
Monotis concentricus, id.	...	Niti							
		Spiti	×	×	×	...	...	...	Genus Monotis.
Avicula echinata, Sow.?	...	Spiti	..	...	...	×	...	Europe	
„ inæquivalvis, Sow.	...	Niti	...	×	...	...	...	Europe	
Pecten æquivalvis, Sow.	..	Spiti	...	×	...	...	...	Europe	
„ comatus, Münst.	...	Niti	...	×	...	...	...	Europe	
„ bifrons, Salter	...	Niti	...	×	...	...	...	...	P. paradoxus, Goldf.
									P. massalonghi, Stopp.
„ monilifer, id.	...	Niti	...	...	×	...	...	...	P. articulatus, Schloth.
„ Lens, Sow.	...	Niti	...	...	×	...	...	Europe	
„ Sabal, Salter	...	Niti							
Ostrea flabelloides, Desh.	...	Niti	...	...	×	...	...	Europe	
„ acuminata, Sow.	...	Niti	...	...	×	×	...	Europe	
Terebratula numismalis, Lam.	...	Niti	...	×	...	...	...	Europe	
„ carinata, Lam. ?	...	Niti	...	...	×	...	...	Europe	
„ globata, Sow.	...	Niti	...	...	×	...	...	Europe	
Rhynchonella variabilis, Schloth.	...	Niti	...	×	...	...	...	Europe	
„ concinna, Sow.	...	Niti	...	...	...	×	×	Europe	
Acrosalenia ?	...	Spiti	...	...	...	...	...	...	The genus in Europe.
Pentacrinites—sp.	...	Spiti	...	×	...	...	...	...	P. basaltiformis, Mill.

Professor H. Blanford, in his very excellent paper lately published in the Asiatic Soc. Journ.,* draws the conclusion from the Spiti lists that the majority of the forms are comparable rather with the Upper Liassic species and those of the Inferior Oolite than any others. The Niti fossils favour the same view, and, except one or two *Ammonites* and the abundant *Rhynchonella concinna*, which may, as Mr. Blanford suggests, have come from the superior beds in the section, there are scarcely any indications of an horizon higher than the one above indicated. We must suppose, therefore, that in this thick series the truly fossiliferous portion belongs to the Lias and Inferior Oolite, while the upper formations are less rich in species, or, quite as likely, have been less successfully searched.

* See Postcript. The original paper, the MSS. from which Mr. Salter quotes, was subsequently much modified.—H. F. B.

Military Orphan Press,—1864.

POSTSCRIPT BY H. F. BLANFORD.

In the interval that has elapsed since the original text of the foregoing pages was prepared, upwards of two years, much has been added to our knowledge of the Palæontology of the Northern Himalya; Professor Oppel of Munich has published a considerable portion of the fossils collected by the Messrs. Schlagintweit, and still more recently, Dr. Ferdinand Stoliczka has visited the Spiti Valley, and has made a far more thorough investigation of the fossiliferous formations of that part of the Himalya than had been accomplished by any previous visitor. I will not attempt here, by any imperfect account of what he has communicated to me, to anticipate the highly interesting additions which Dr. Stoliczka has made to our knowledge. Being an experienced Palæontologist, he has brought his knowledge of fossil forms to bear on the study of the Spiti formations *in situ*, and the result has been that he has defined their stratigraphic and palæontological relations with a completeness far surpassing anything previously effected. Dr. Stoliczka is now engaged in working out the fossils collected by himself, as well as those previously obtained by Mr. Theobald, and I do not doubt that his more extensive material and trained judgment will enable him to correct my own work on the Jurassic fossils of the Himalya on several points. These, however, I shall leave to him to point out in his forthcoming work, and shall content myself in this postscript with making such corrections as are necessitated by the prior appearance of Professor Oppel's work on the Jurassic ammonites, a copy of which that gentleman has most kindly furnished to me. Professor Oppel's nomenclature will of course take precedence of that of the foregoing pages. It is, however, of later publication (1864) than that in text of my paper in the Journal of the Asiatic Society (1863).

The following parallel lists represent what now appear to be synonyms; to prevent future confusion, I have marked with an (*) those which claim priority of publication:—

Blanford, 1863.	Prof. Oppel, 1864.	Blanford, 1865.
* *A. octagonus*, Str.	*A. Sömmerringi*, Opp.	{ *A. octagonus*, Str. { *A. Hookeri*, Str.
	* *A. Stanleyi*, Opp.	*A. scriptus*, Str.
* *A. Spitiensis*, Blanf.	*A Groteanus*, Opp.	*A. Spitiensis*, Blanf.
* *A. guttatus*, Str.	*A. Cautleyi*, Opp.	*A. guttatus*, Str.
* *A. Gerardi*, Blanf.	*A. cognatus*, Opp.	
* *A. Wallichii*, Gray.	*A. Morikeanus*, Opp.	*A. Wallichii*, Gray.
	* *A. Sabineanus*, Opp.	*A. Jubar*, Str.
* *A. acucinctus*, Str.	*A. Lymani*, Opp.	*A. acucinctus*, Str.
	* *A. Theodorii*, Opp.	*A. Griffithii*, Str.
* *A. (Cer.) Himalayanus*, Blanf.	*A. Thuillieri*, Opp.	
* *A. Hyphasis*, Blanf.	*A. Seideli*, Opp.	*A. umbo*, Str.
* *Belemnites sulcatus*, Mill.	*B. Gerardi*, Opp.	*B. sulcatus*, Miller.
* *Astarte major*, Sow.	*A. Hermannii*, Opp.	

For some of the above identifications, I am indebted to Dr. Stoliczka.

As mentioned at page 75, several references to my paper on Dr. Gerard's fossils have been corrected in this reprint. This paper, as originally written, had described under other names several of the species which exist in Colonel Strachey's collection, and among these, some, which I had at first regarded as identical with European species, proved, on examination of Colonel Strachey's more perfect specimens, and of authentic specimens of European species, to be distinct. The original manuscript had been left by me for publication in the hands of the Asiatic Society of Bengal before I left India, and contemplating the possible event of the paper being published during my absence in England, in describing Colonel Strachey's fossils, I quoted my own previous names of identical species as synonyms, mentioning at the same time in the text, the errors which I had made in my paper on Dr. Gerard's fossils. On my return to India, I found that my paper had been reserved for my revision, and I accordingly withdrew and corrected it to the form in which it appeared in the

Journal of the Asiatic Society, 1863. Mr. Salter's quotations of my paper are taken from a copy of the original manuscript which I had with me in England, and there are one or two points in these quotations which I have felt it better not to alter in the reprint, but which require explanation, as they are at variance with the views enunciated in my paper as published.

The chief of these is the reference to the Liassic species *A. communis, A. heterophyllus, A. Thouarsensis, A. bifrons, A. concavus* and *Pecten æquivalvis,* which will be found enumerated in the list of Himalyan fossils at page 102, and five of which were figured by Mr. Prinsep as forming part of Dr. Gerard's collection (As. Res. Vol. XVIII). This latter authority I had unhesitatingly accepted as evidence that the species in question formed part of Dr. Gerard's Spiti collection, nor had I any reason to suspect otherwise, until Mr. Theobald, on returning from a visit to Spiti, informed me that he had obtained no fossils similar to them, and pointed out that none of the undoubtedly genuine fossils had a similar matrix. I have already (Journ. As. Soc., Vol. XXXII., p. 124) published my reasons for the conclusion at which I eventually arrived, viz., that the species in question were really Whitby fossils that had been accidentally mixed up with Dr. Gerard's collection, a conclusion which has been disputed by Mr. Oldham (Journ. As. Soc., Vol. XXXIII, p. 237), but which, on careful re-consideration, I see no reason to alter. Dr. Stoliczka's examination of the Spiti Valley and its neighbourhood has been made subsequently to the reading of Mr. Oldham's paper, and he was well aware of the question at issue. He has examined the country more thoroughly than any of his predecessors, as his published report will doubtless show, and he was especially careful to re-visit and examine all the localities mentioned by Dr. Gerard; but, although he has obtained fossils which appear to be of Liassic age, he has not found any of the species above enumerated, nor have his Liassic fossils the characteristic 'Whitby' matrix.

On the other hand, Dr. Oppel describes and gives a figure of an Ammonite from Gnari Khorsum, in M. Schlagintweit's collection, *A. Kobelli,* which is possibly only a variety of *A. bifrons.* The matrix of this is described as a black geode, while that of Dr. Gerard's (questionable) specimens is a blue calcareous shale, identical with that of authentic Whitby fossils.

Mr. Oldham, in the paper referred to above, adduces the existence of *A. communis,* &c., in the Oxford Museum, as independent evidence to the authenticity of the fossils rejected by me, an aspect in which I am unable to accept them, unless it can be proved that these fossils (which are stated to have been obtained by Dr. Gerard) were forwarded to England at a time or under circumstances which preclude the possibility of the intermixture having taken place in the Asiatic Society's Museum or elsewhere, before the specimens were despatched. This important point appears to have been overlooked by Mr. Oldham in his published paper, and I accordingly endeavoured, through the kind agency of my friend Professor Maskelyne, to obtain some information upon it. Professor Phillips' note which I received in reply is given below;* I should infer from this that Professor Phillips is by

* "*Notes on Himalayan Fossils in the Museum at Oxford; June 2nd,* 1864.
By Prof. John Phillips.

"About 30 years since I sent from York to Calcutta a considerable series of the fossils of Whitby and some other tracts. The specimens were selected from the duplicates of the Yorkshire Philosophical Society, and were presented by that institution to some individual of position in Calcutta, whose name I cannot remember (it *seems* to me to have been Patterson), but could find out. Whether the collection was carefully kept separate at Calcutta, I know not; but some years afterwards, on being shown in England a specimen of *A. communis,* said to be 'from the Himalaya,' I at once conjectured that the Yorkshire collection *might* have given forth this offset, so like—so absolutely like—in form, colour, and accompaniments of shale or ironstone. The same astonishing resemblance occurs in regard to these specimens in the Oxford Museum, especially in regard to the *Ammonites communis* and *A. bifrons* (*Walcottii*), which are very common at Whitby.

"On the other hand, the other fossils in this collection do not offer any especial analogy to Yorkshire types; some are of Oxfordian shapes, and of the Belemnite in particular, it is unknown in Yorkshire, but a good deal like some found in the South of England, *as to form;* not, I think, as to conservation, &c.

"Among the fossils we note as of Liassic age, *Ammonites communis.*

no means convinced of the genuineness of some of the fossils in the Oxford Museum, but the note throws no light on the circumstances and dates of their transmission from India.

Now, however, that Ammonites of distinctly Liassic affinities, the authenticity of which cannot be questioned, have been obtained from Gnari Khorsum by Mr. Schlagintweit, and that Dr. Stoliczka has obtained Liassic fossils from Spiti, the question of the genuineness of certain specimens in Dr. Gerard's collection has lost very much of its interest, and is restricted to the authenticity of certain specimens in the Calcutta and Oxford Museums, and to the inference to be drawn therefrom of the occurrence of the four or five disputed species in the Liassic rocks of the North Himalya. These do not appear to me of sufficient importance to warrant the prolongation of the discussion to which the fossils originally

---

"*Ammonites bifrons* (*Walcottii*); such occur at Whitby. The variety of *A. communis* called *crassus*, is found both at Whitby, and in this series, from the Himalaya!

"*Pachyodon Listeri* in plenty. It is not quite like ordinary English specimens.

"Small *Spirifera* of the Liassic type, such as occurs in South of England, not in Yorkshire.

"With this *Spirifera* in *plenty*, occur:—

"*Rhynchonella* of the types *concinna* and *obsoleta*. In separate masses.

"*Avicula* like *Braamburiensis*.

"*Astarte*.

"*Trigonia* of a type near middle and top of Bath Oolite series, not quite like any English form, and separate.

"*Belemnites* of the group *B. sulcatus*, Miller, probably of Oxford clay.

"Palœozoic Fossils also occur, including,—

"*Prod. antiquata*.

"*Spirifer* 1.
2.
3. *attenuata*

"*Strophomena*.

(Sd.) 'JOHN PHILLIPS,'
Oxford.

"2nd June 1864."

gave rise, and I shall be content to wait for any further evidence which time may disclose.*

The following Note by Dr. Oldham has been handed to me by Colonel Strachey. It is but just to Mr. Theobald to observe, that unless Dr. Oldham has privately communicated the Note to him, of which I am not aware, he has had no opportunity of knowing and replying to Mr. Oldham's remarks.

"In the preceding pages, several references are made by Mr. Salter (pp. 51, 54, 58) to a small part of the collection of fossils from the Spiti Valley, belonging to the Geological Survey

---

* As the present publication will terminate my part in the discussion, I will describe at more length than I have hitherto done, the circumstances under which Dr. Gerard's fossils were found by me in the Asiatic Society's Musuem. The specimens figured in the Asiatic Researches were (except in one or two cases in which specimens were labelled in ink as from the Himalya,) my only guides to the genuineness of the collection. These were found partly on the shelves of a small wall case, partly in an old box without label, the other contents of which were a large number of *A. communis* and *A. bifrons* with a few of *A. concavus* and *A. Thouarsensis*, also some other specimens of *Ammonites, &c.*, which last, although unlabelled and unfigured, the peculiar black geode matrix and conservation indicate, with little room for doubt, as from the Himalya. All these were mixed indiscriminately. They bore no label and the wrapping paper (if there had ever been any around them,) had entirely disappeared. There were no other fossils in the box (so far as I remember); certainly no other Whitby species than those figured in the Asiatic Researches and *A. Thouarsensis*. There did not seem, therefore, to be any *á priori* reason for doubting that the specimens in the box formed the bulk of Dr. Gerard's collection, from which a few specimens (not by any means all of those figured) had been put out in the wall case. Since the publication of my paper, in hunting through the Asiatic Museum, I have found a large additional number of fossils. which, from their specific and lithologic identity with [now] described Himalyan fossils, I infer to have also formed part of Dr. Gerard's collection. These were partly in a case of miscellania, (mixed with European and Australian fossils &c ,) partly in a box mixed up with Palæosoic fossils from Van Dieman's Land and New South Wales. I have not, however, described these, as Dr. Stoliczka is engaged on the description of his more perfect series, and he will utilize the Society's fossils so far as they are of any value.

There is some difficulty in recognizing the originals of certain of Mr. Everest's figures in the As. Res., but those of *A. concavus*, fig. 7, *A. heterophyllus*, fig. 9, and *Pecten æquivalvis*, fig. 20, have been satisfactorily determined. There are two or three specimens, any one of which might have been the original of *A. communis*, fig. 5, and of *A. bifrons*, fig. 6.

Museum, Calcutta, which I brought to London during a hasty visit in 1862, for comparison with Colonel Strachey's and other specimens. This collection had been made in 1861 by Messrs. Theobald and Mallet. In going over these with my friend Mr. Salter, I stated to him that, so far as represented by Mr. Theobald, there were only two recognizable groups among these fossiliferous rocks, one lower, one upper, (excluding the silurian), and that the lower, according to him, held the Spirifer described above as *Spirifer Rajah* (which we had referred to *S. Keilhavii*, and which further comparison of other specimens has not enabled us to separate from VonBuch's species), and the ammonites which we had recognized as Triassic. As I had never been able to visit the locality, I could only give the information collected by others. Mr. Theobald's statements were very distinct on this point, though not supported by Mr. Mallet. The country is, however, difficult, and I felt convinced that some error had crept in. Under this conviction, I did not publish the results obtained in 1861, and in consequence, I despatched again to the same localities, in the present year, Dr. F. Stoliczka and Mr. Mallet, and with the most satisfactory results. These will be published in more detail hereafter, but in connection with the preceding pages, I may state that the *Spirifer Rajah* (of Salter) does *not* occur (see *ante*, page 54) in the same beds with Triassic ammonites, but in beds decidedly below them,—beds which other evidence combines to show must be referred to the same general relative age (in the sense of *homotaxis*) as the 'Carboniferous' of Europe."

ERRATA.

Page 78, line 3, for *A. tenuisulcatus* read *A. tenuistriatus*.
„ 87, „ 8, „ ACUCINTUS read ACUCINCTUS.
„ 101, „ 18, „ *P. basaitiformis* read *P. basaltiformis*.
„ 102, Erase *A. Eugenii*. Rasp. from list.

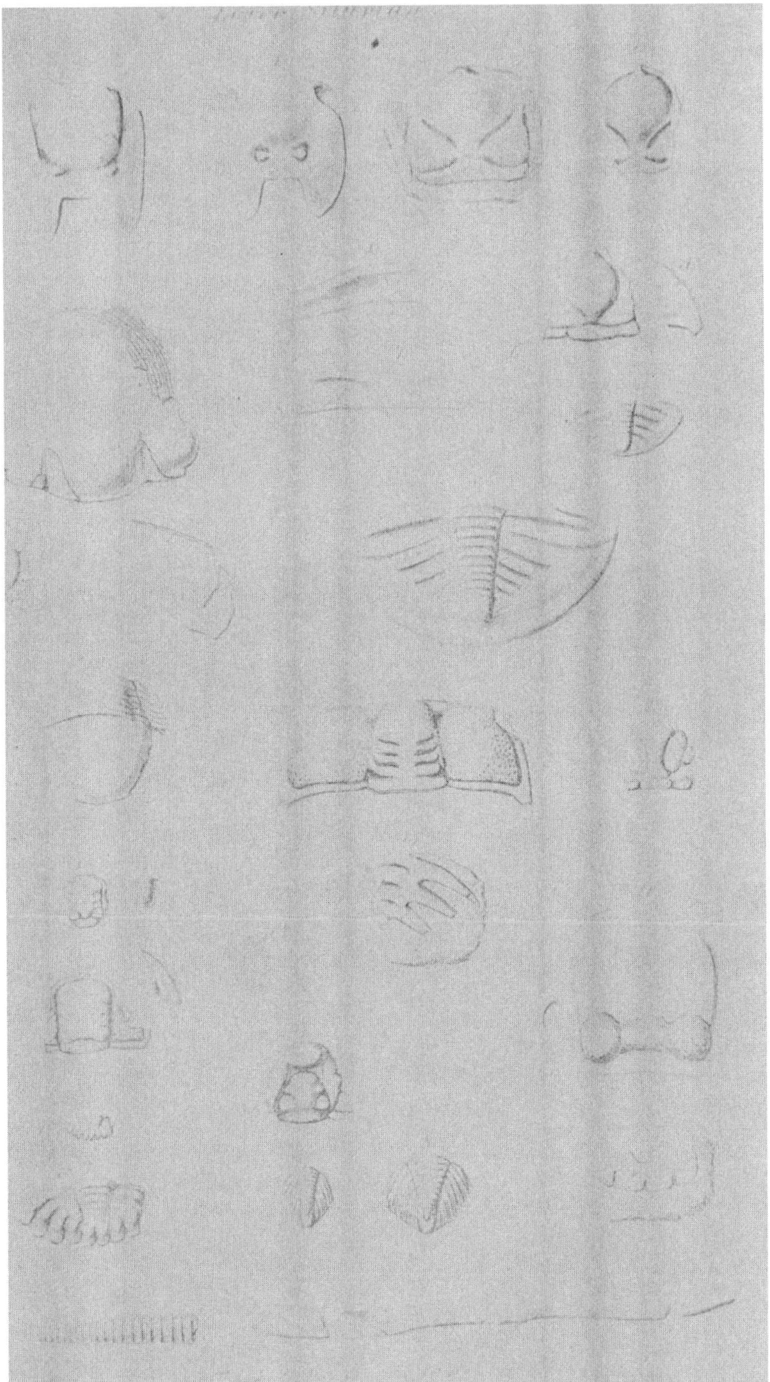

Lower Silurian

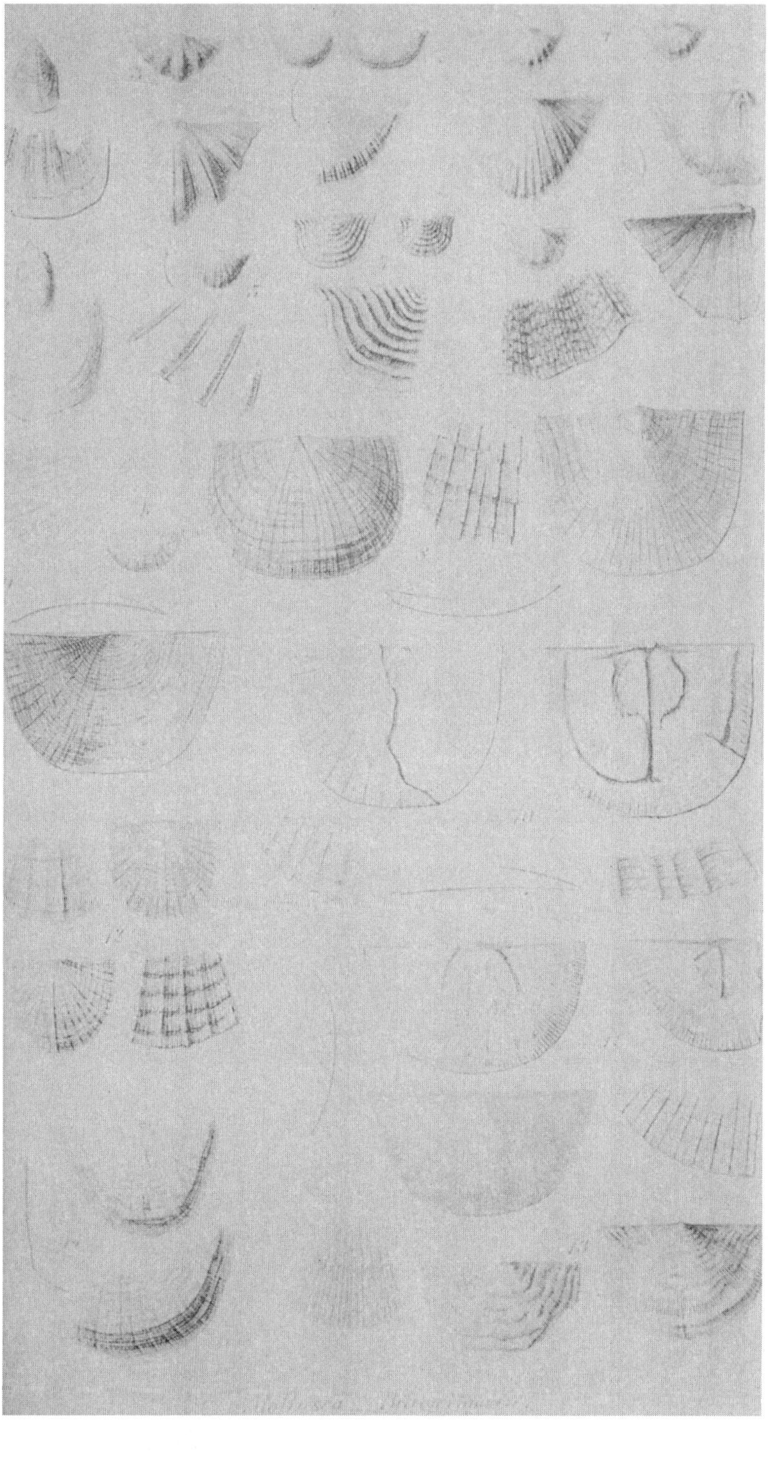

Lower Silurian

Brachiopods and Polyzoa

J.W. Salter del.

J.D.C. Sowerby Sculp.

Carboniferous (Brachiopods).

Upper Triassic (Hallstatt Beds) Vol. II Pl.

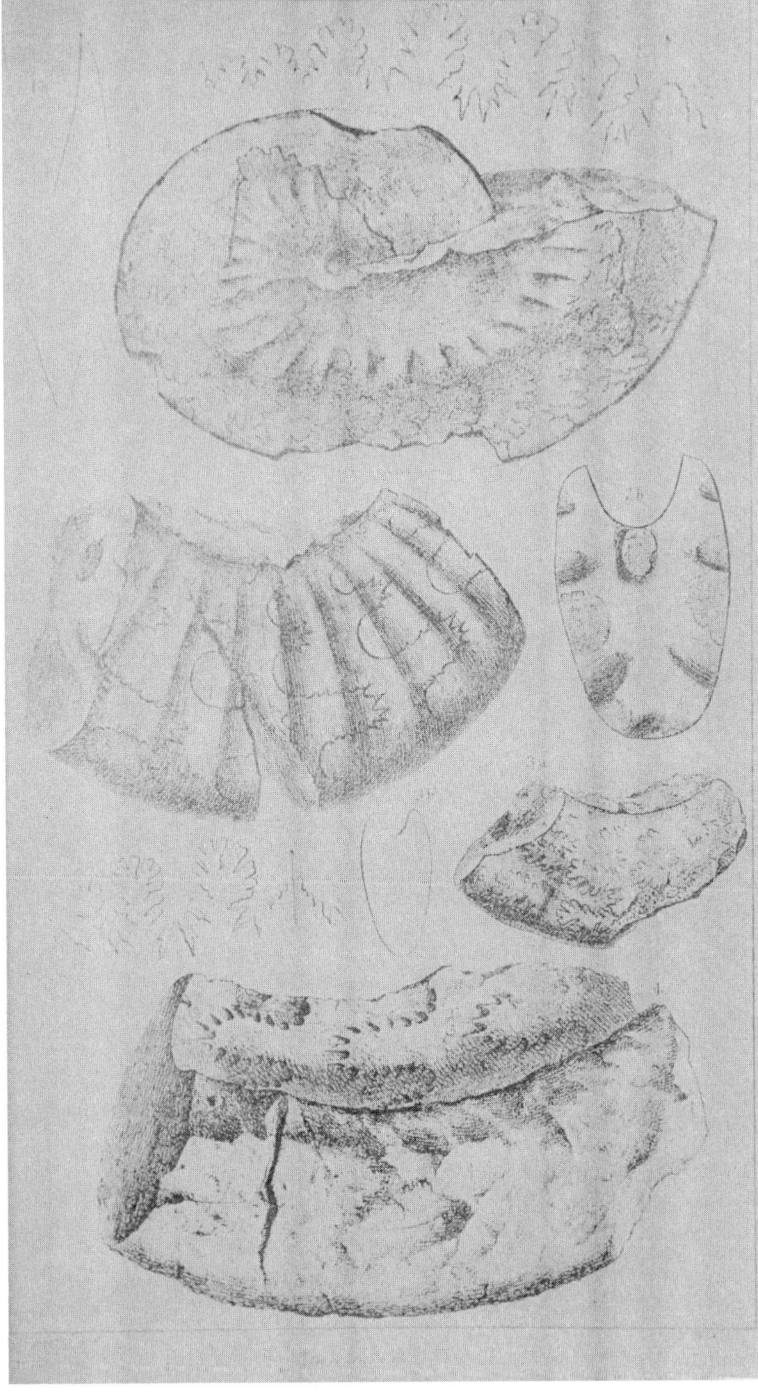

Upper Triassic (Dachstein beds)

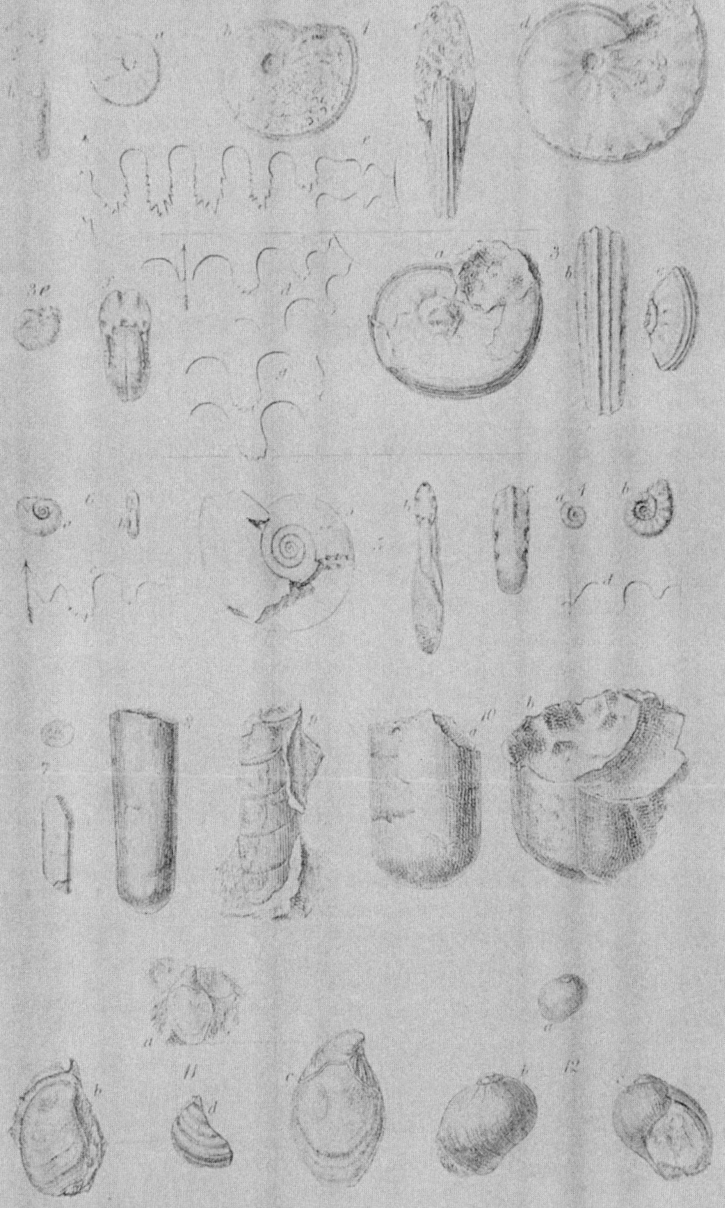

J. D. C. Sowerby fecit.

Mollusca _ Univalves.

V.III Pl.9

Upper Triassic (Dachstein beds)

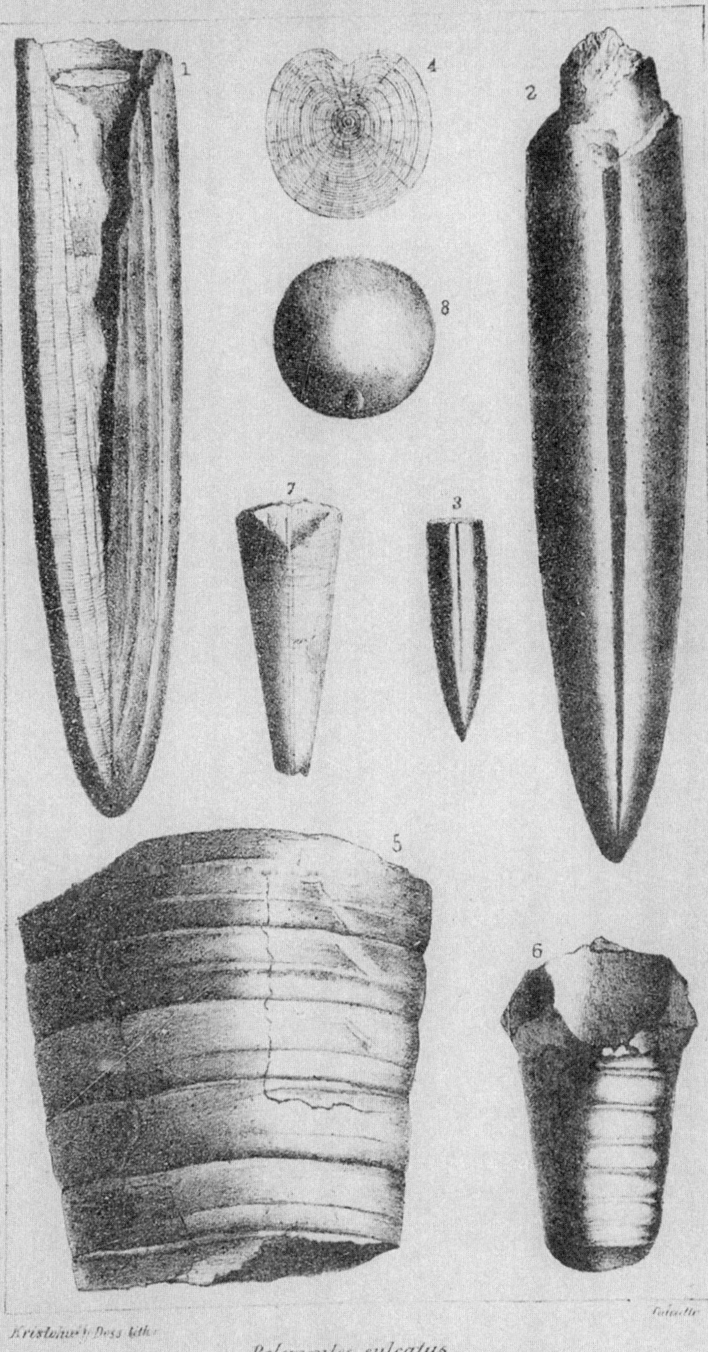

Belemnites sulcatus.

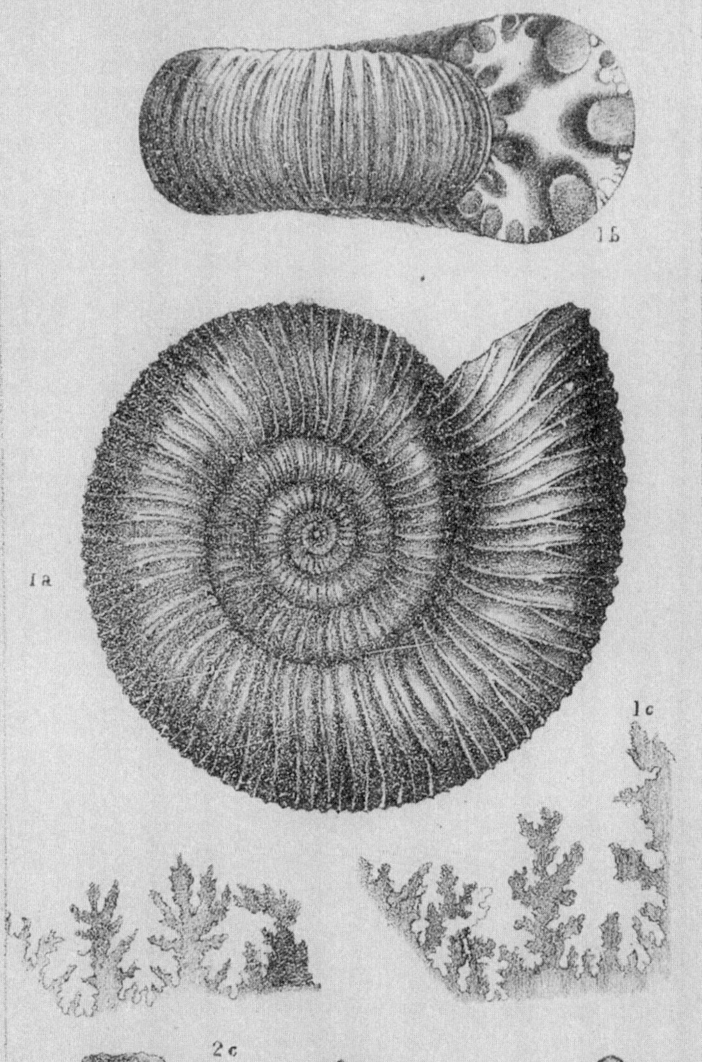

Kristohury Doss lith. Calcutta

1. Amm. biplex. 2. A. Batteni.

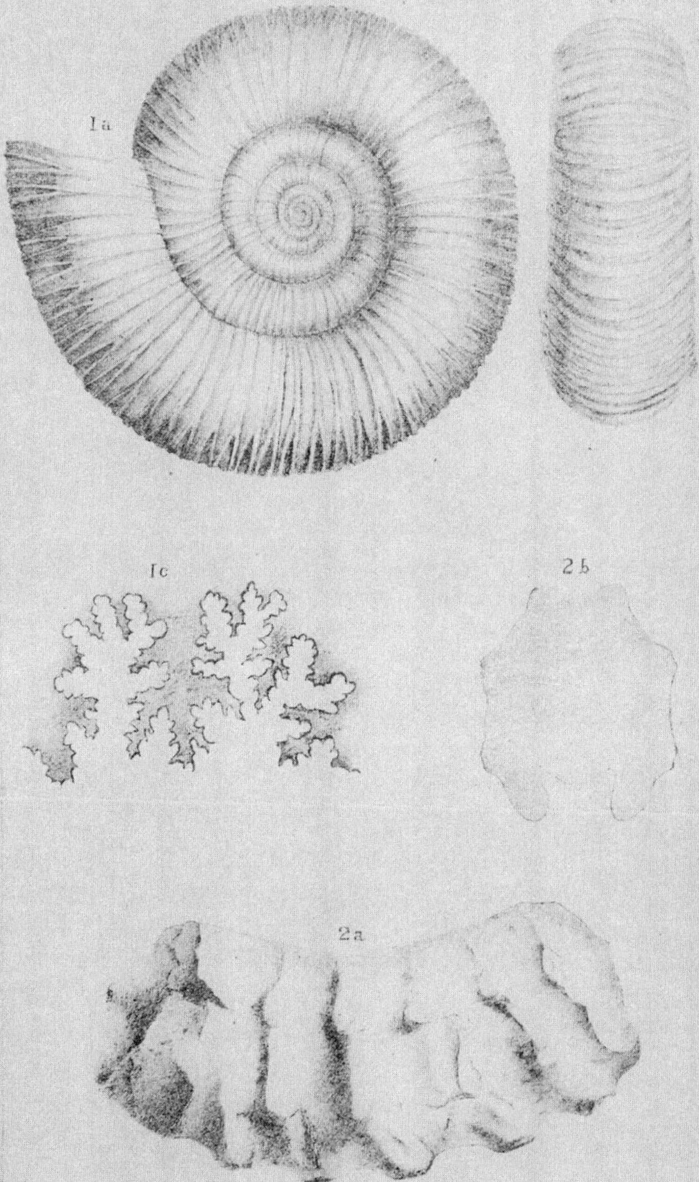

Kristobury Dess. lith.

Calcutta

1. Amm. biplex. 2. A. octagonus.

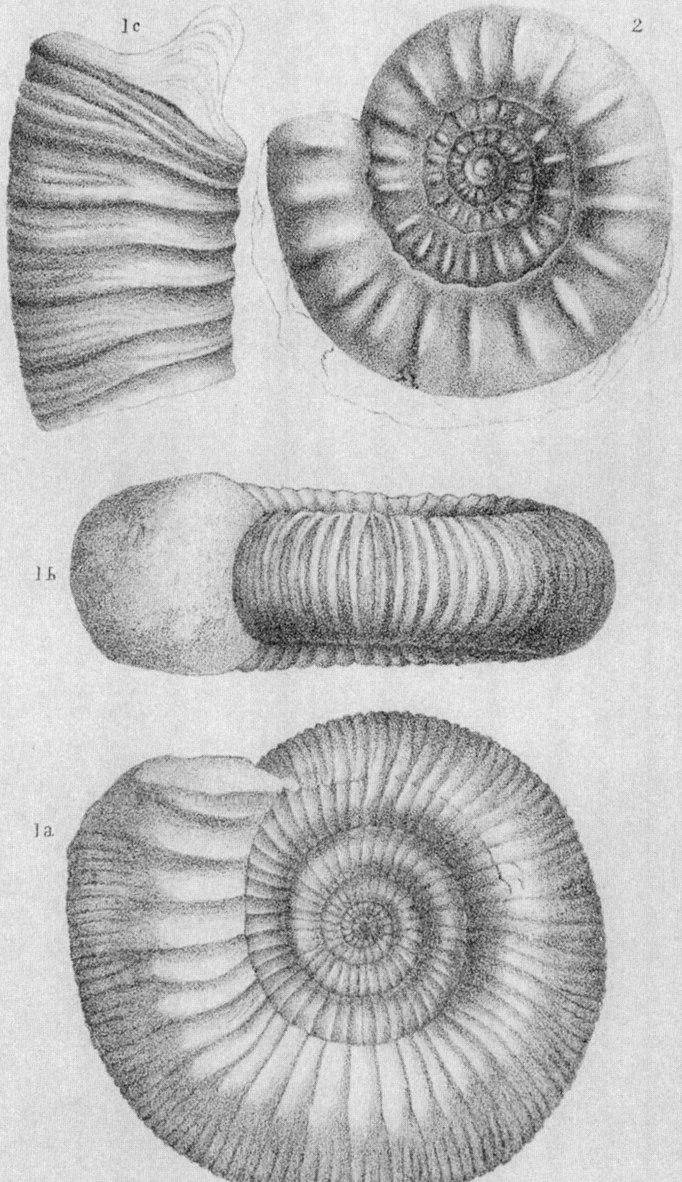

Kristohury Doss lith. *Calcutta*

1 Amm. triplicatus 2 A. guttatus

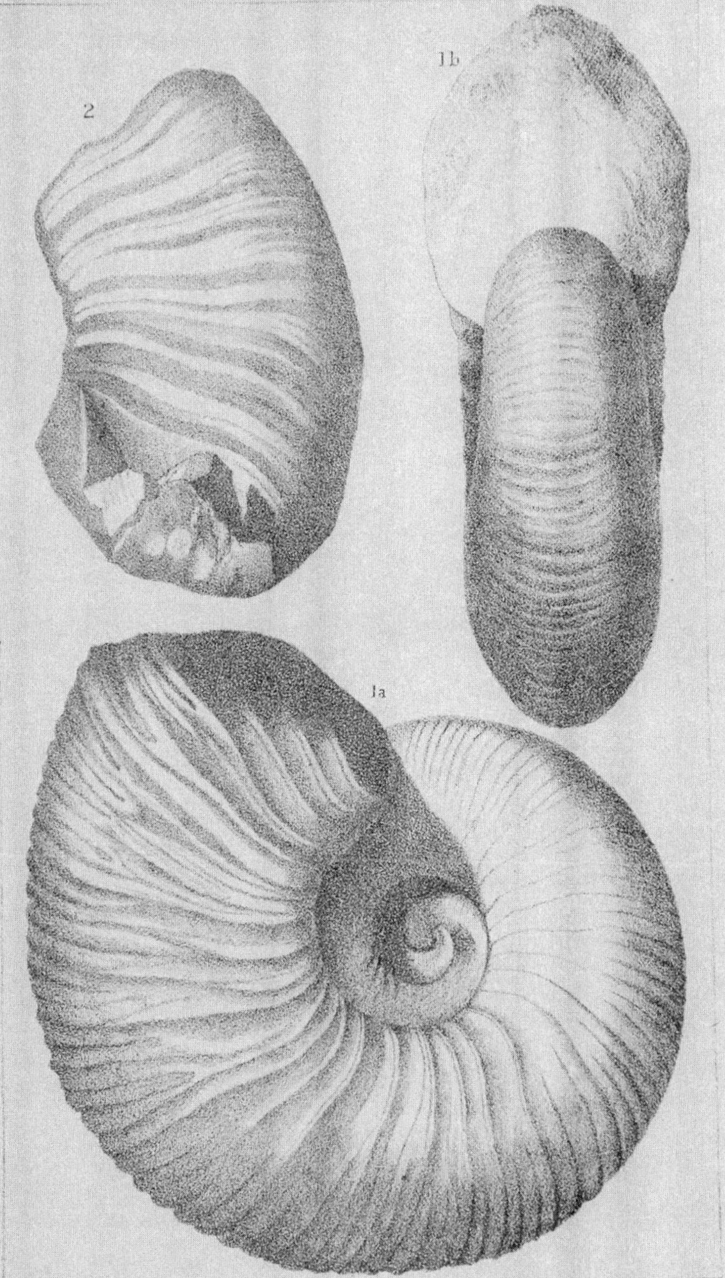

Kristohury Doss lith. Calcutta

1. A. Nepalensis. 2. A. tenuistriatus.

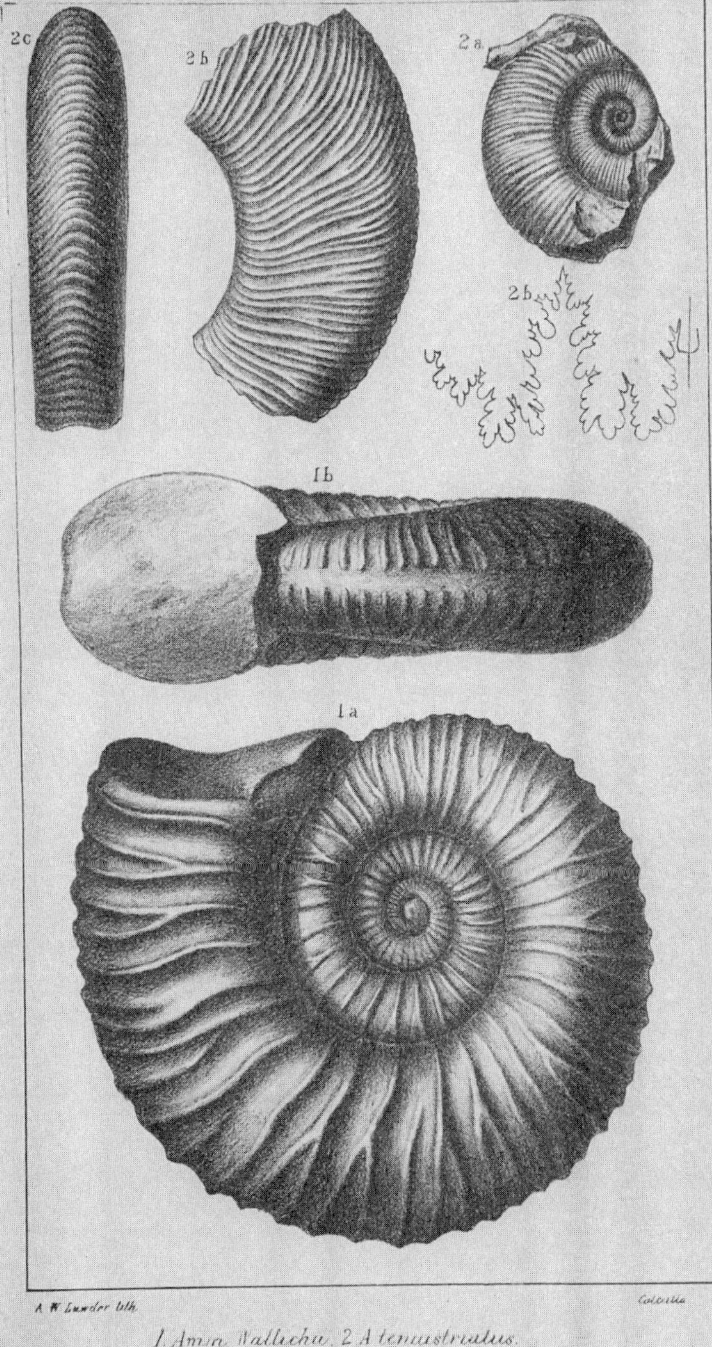

A. W. Lunder lith. Calcutta

1. Amm. Wallichii. 2. A. tenuistriatus.

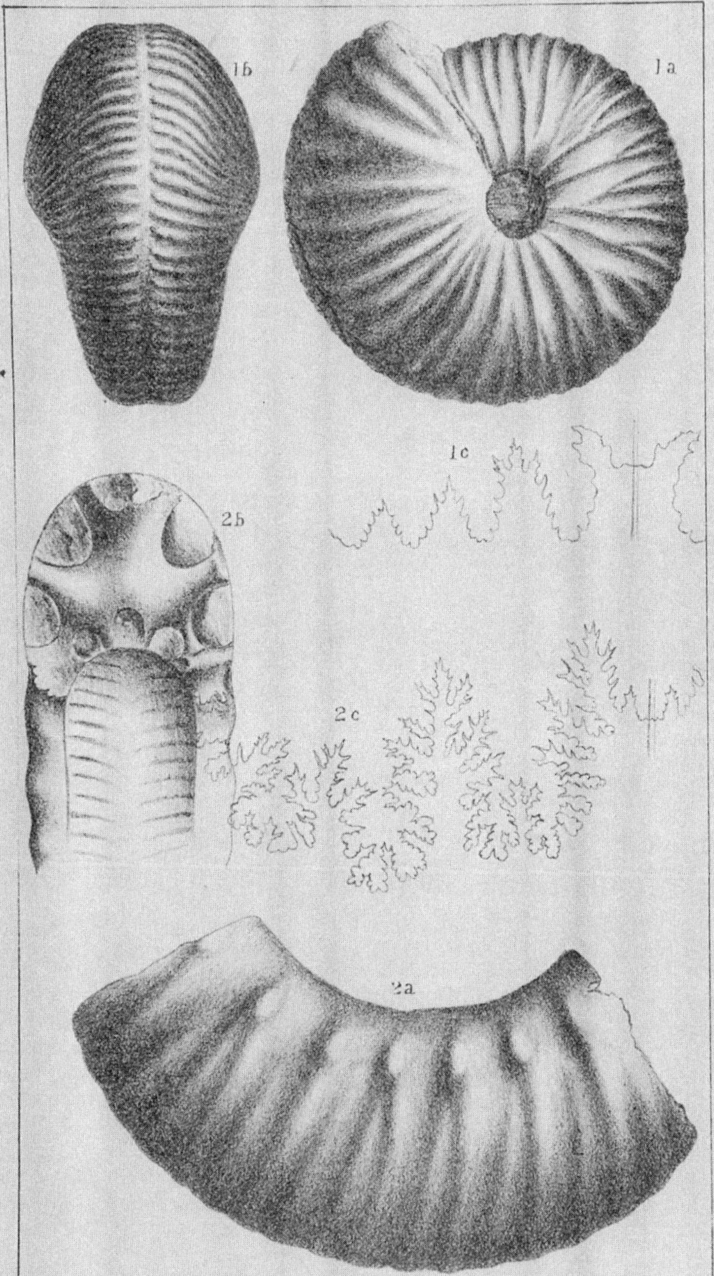

1 Ammonites robustus 2 A. scriptus.

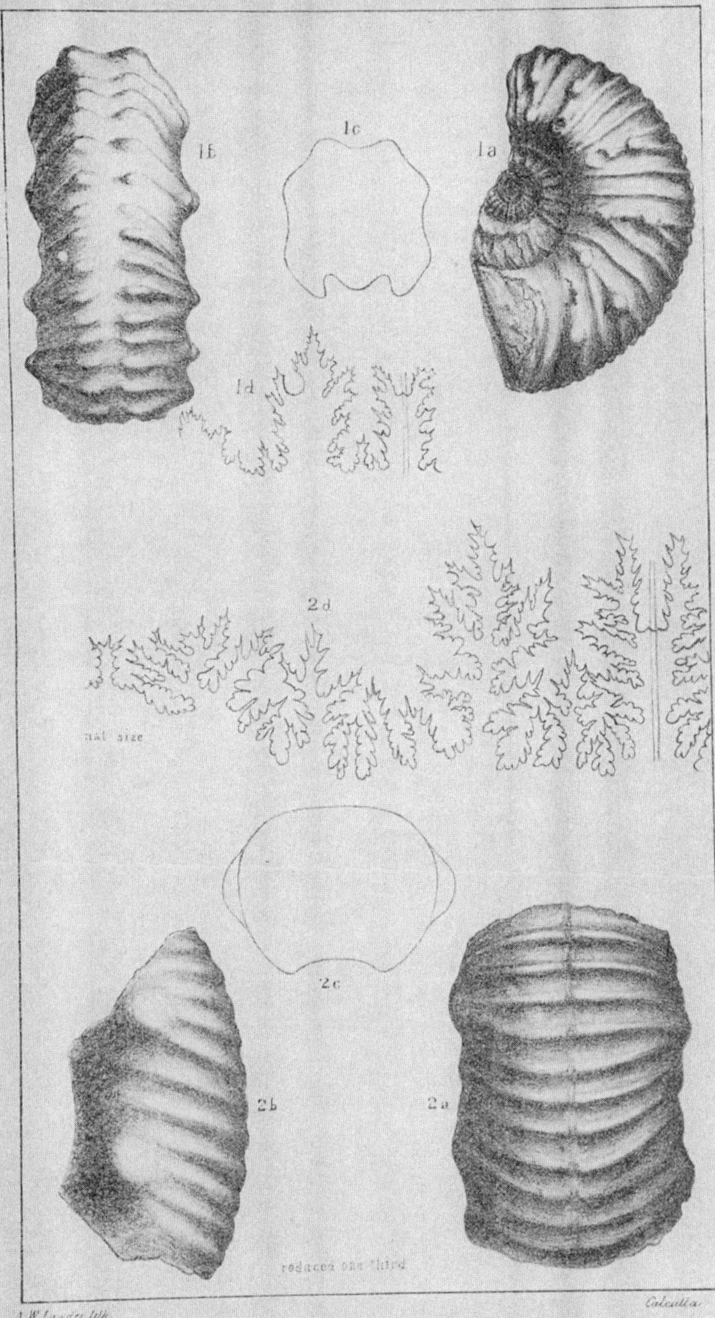

A. W. Lander lith. Calcutta

1 Amm. Hookeri 2 A. umbo.

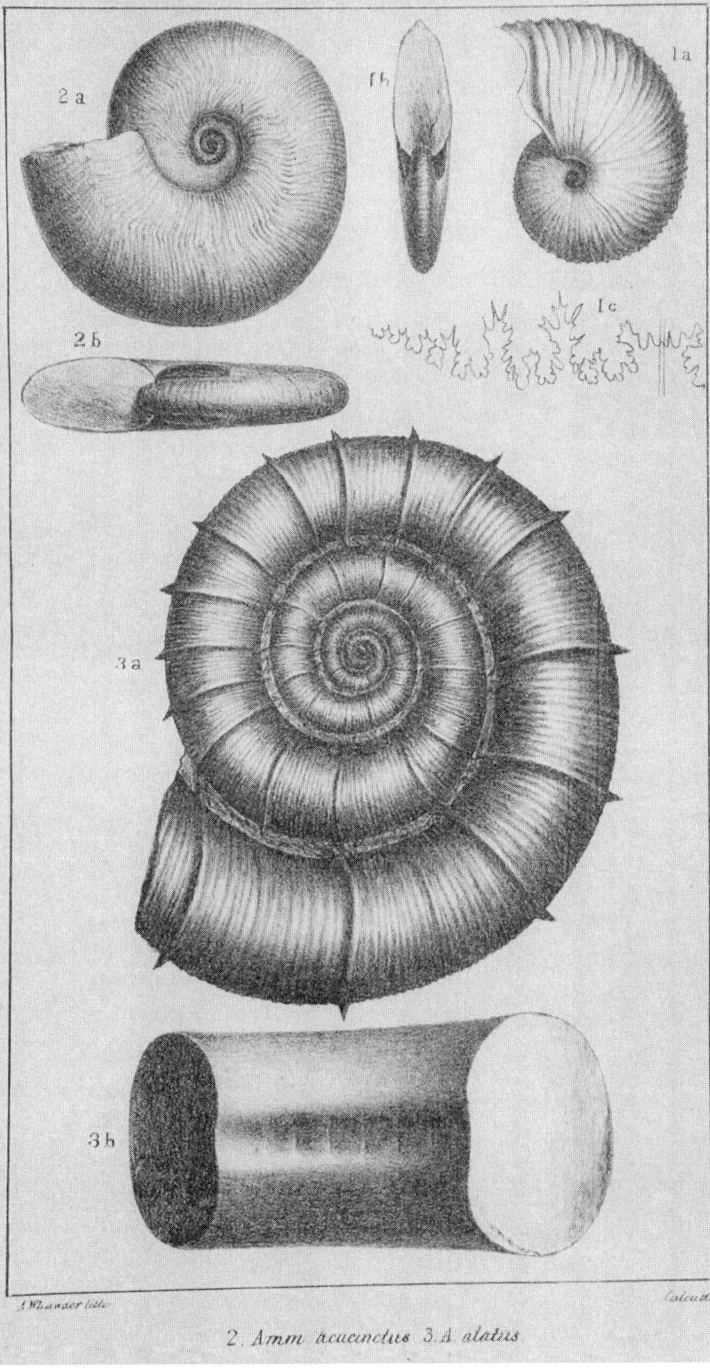

J. Whasser lith. Calcutta

2. Amm. acacinctus 3. A. alatus

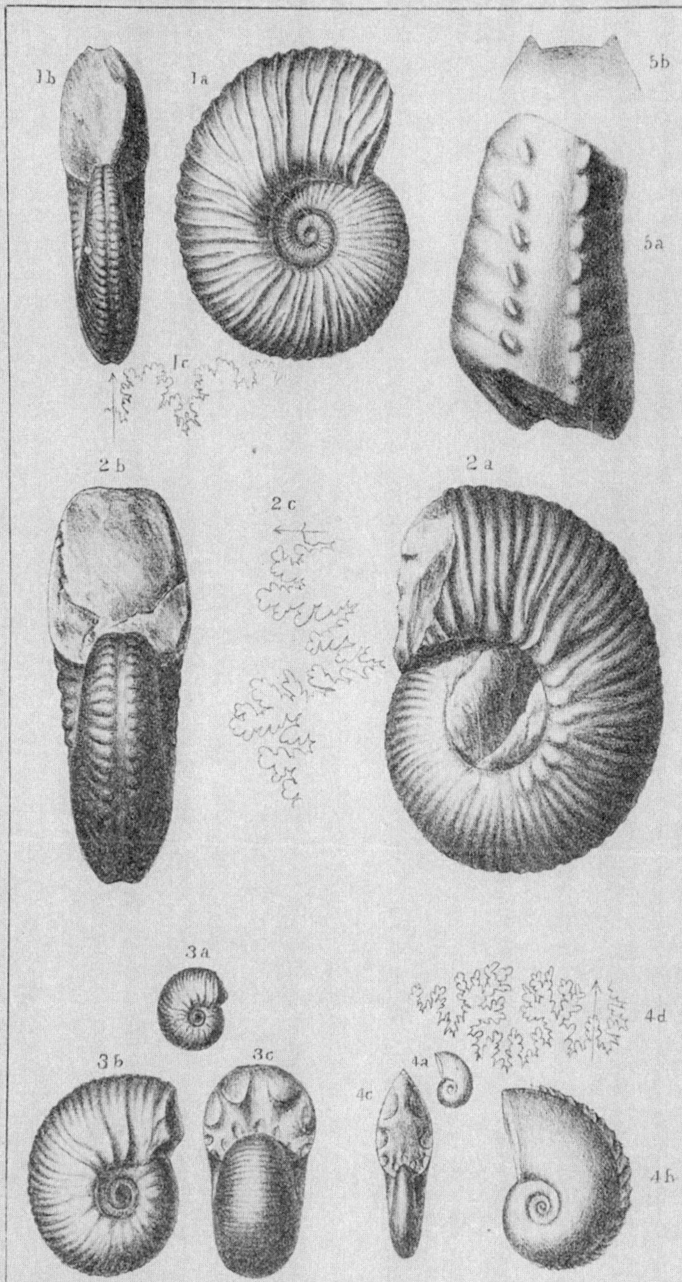

Calcutta

Ammonites.

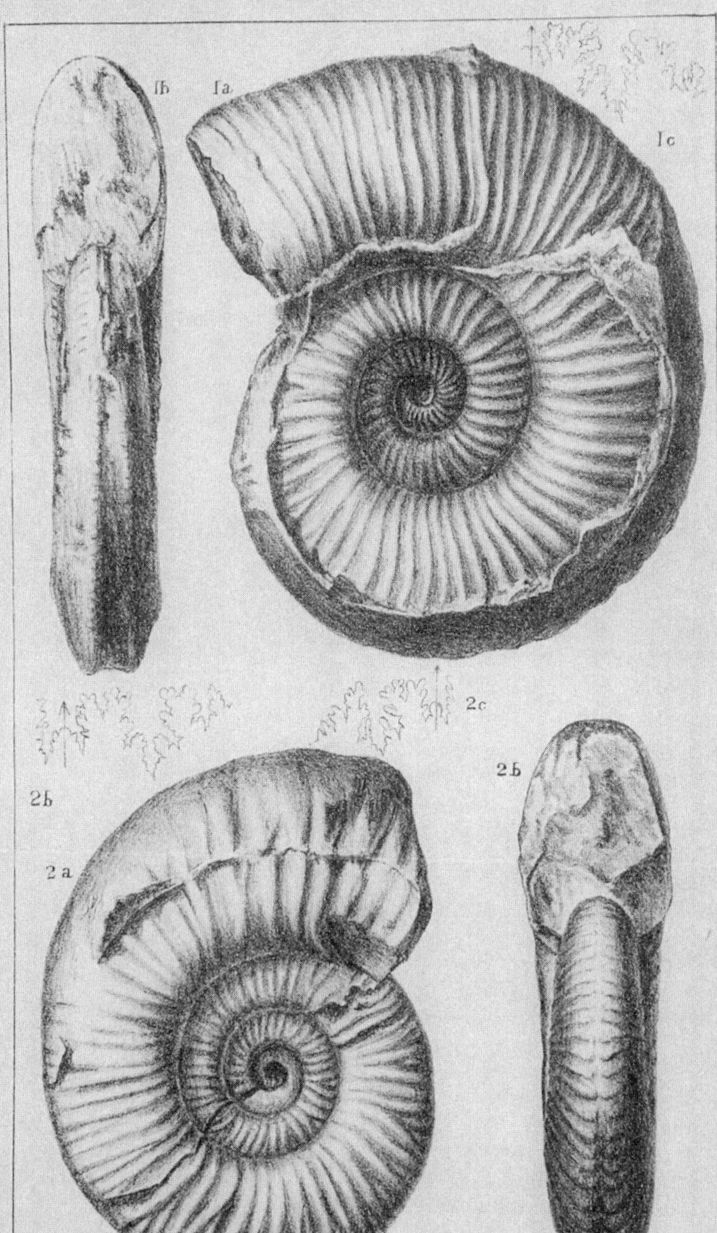

Calcutta

1. *Amm. Griffithii* 2. *A. jubar*

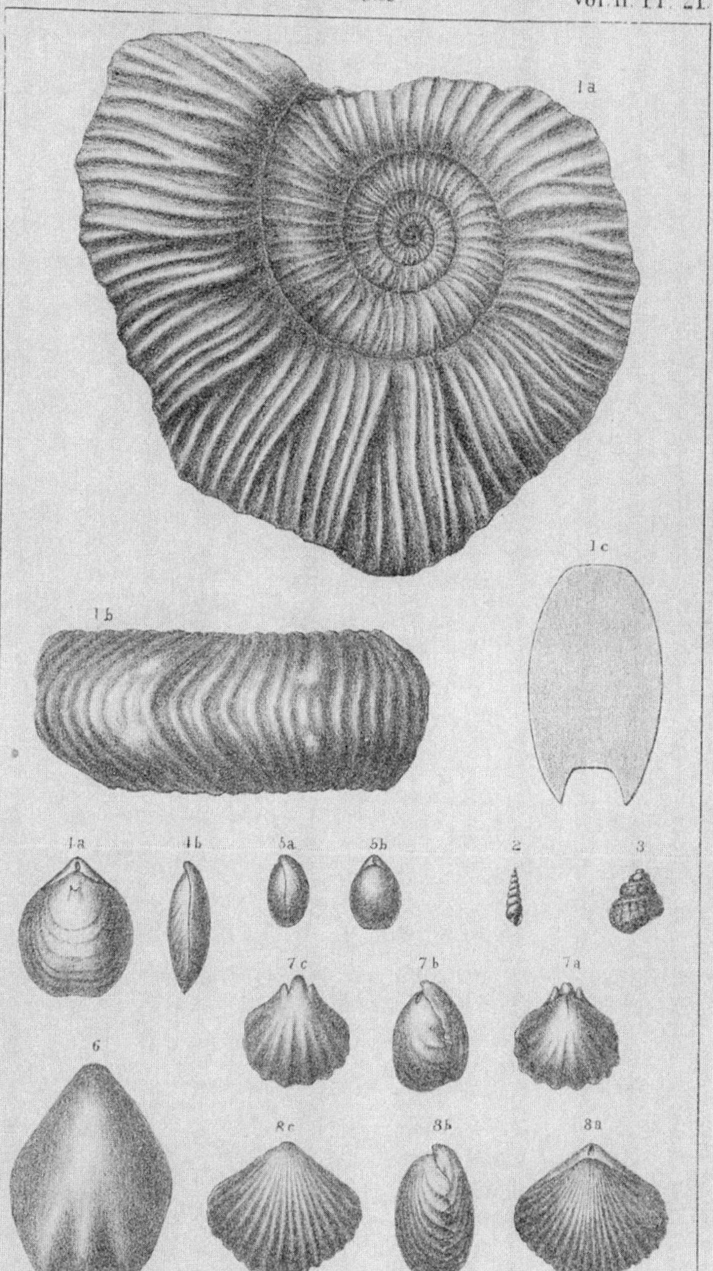

Ammonites & Brachiopoda.

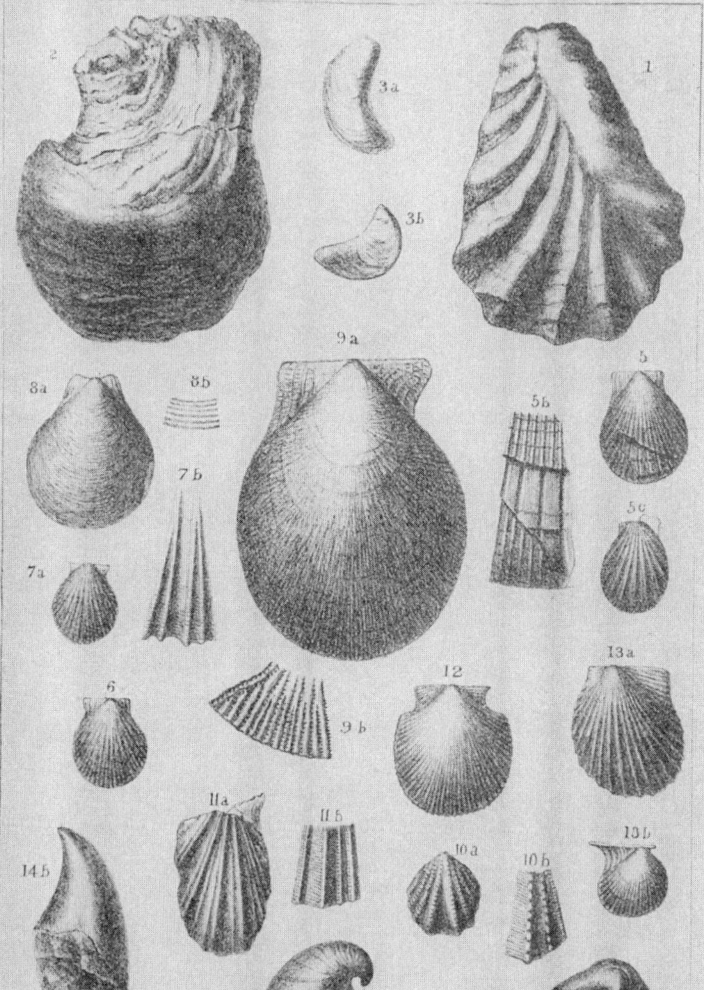

A. W. Lowder lith. Calcutta

Lamellibranchiata Monomyaria

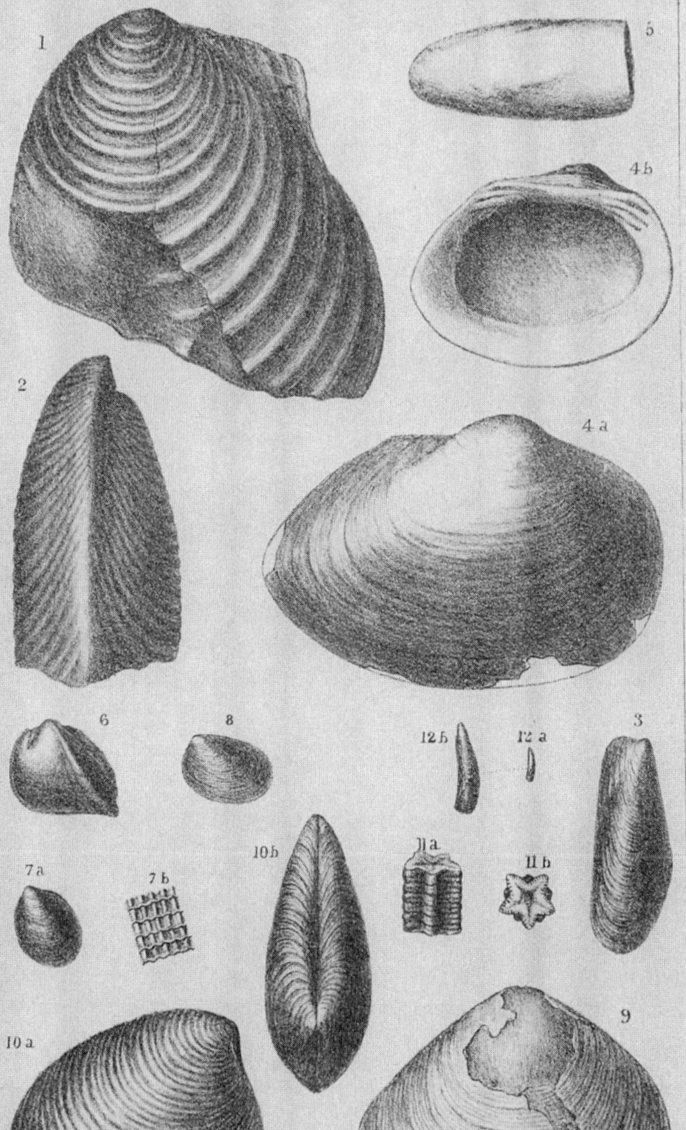

A. W. Lander lith.

Lamellibranchiata, &c.